ELEMENTS, PRINCIPLES AND CORPUSCLES

ARCHIVES INTERNATIONALES D'HISTOIRE DES IDÉES

INTERNATIONAL ARCHIVES OF THE HISTORY OF IDEAS

171

ELEMENTS, PRINCIPLES AND CORPUSCLES

by
ANTONIO CLERICUZIO

ANTONIO CLERICUZIO

ELEMENTS, PRINCIPLES AND CORPUSCLES

A Study of Atomism and Chemistry in the Seventeenth Century

KLUWER ACADEMIC PUBLISHERS
DORDRECHT / BOSTON / LONDON

A C.I.P. Catalogue record for this book is available from the Library of Congress.

ISBN 0-7923-6782-0

Published by Kluwer Academic Publishers,
P.O. Box 17, 3300 AA Dordrecht, The Netherlands.

Sold and distributed in North, Central and South America
by Kluwer Academic Publishers,
101 Philip Drive, Norwell, MA 02061, U.S.A.

In all other countries, sold and distributed
by Kluwer Academic Publishers,
P.O. Box 322, 3300 AH Dordrecht, The Netherlands.

Printed on acid-free paper

Printed in the Netherlands.

To Rita and Micol

CONTENTS

PREFACE

I became interested in the relationship between chemistry and corpuscular philosophy in 1983, while writing my Dissertation on Robert Boyle under the supervision of Professor Tullio Gregory at the University of Rome, 'La Sapienza'. Thanks to a three-year Frances A. Yates fellowship at the Warburg Institute, I was able to investigate seventeenth-century English chemistry and medicine, as well as the unpublished manuscripts and letters of Robert Boyle. This work made it apparent to me that the widely accepted notion of mechanical philosophy was too comprehensive and too vague. Hence, my aim was to understand the variety of versions of the corpuscular philosophy and in addition to assess the complex relationship between chemistry and atomism. I have expounded my views on Boyle and seventeenth-century chemistry in various articles and seminars, which have been extremely useful for clarifying my interpretation of the corpuscular philosophy.

In the several years in which this book has been in preparation I have accumulated many debts of gratitude. It is a pleasure to express here my thanks to Charles Webster for his intellectual and moral encouragement since my arrival in England in 1985. I am also grateful to Simon Ditchfield, Michael Hunter, Didier Kahn, Christoph Lüthy, Roberto Palaia and Michela Pereira for their information and comments. Special thanks are due to Mrs Constance Blackwell for her advice and support. I am also grateful to Pietro Corsi for his encouragement.

My thanks are due to the Warburg Institute, the Foundation for Intellectual History, the University of Cassino and the Consiglio Nazionale delle Ricerche for support during my research in London.

The staff of the following libraries deserve special thanks: the British Library (London), Library of the Warburg Institute (London), the Wellcome Library (London), the Bodleian Library (Oxford). For permission to quote from the Boyle Papers I am grateful to the Council of the Royal Society. I also thank the Permanent Secretaries of the Académie des Sciences (Institut de France) for permission to quote from the manuscripts of the Archives of the Académie.

Rome, September 2000 A.C.

ABBREVIATIONS

AT	R. Descartes, *Oeuvres,* ed. C. Adam and P. Tannery, 12 vols. (Paris, 1897-1913)
BL	British Library
BP	Royal Society Boyle Papers
DBI	*Dizionario Biografico degli italiani*
DNB	*Dictionary of National Biography*
DSB	*The Dictionary of Scientific Biography*, ed. C.C. Gillispie 16 vols. (New York, 1970-80)
Ferguson	J. Ferguson, *Bibliotheca Chemica: a Catalogue of the Alchemical, Chemical and Pharmaceutical Books in the Collection of the late James Young* 2 vols. (Glasgow, 1906)
Frank, *Harvey*	R.G. Frank jr., *Harvey and the Oxford Physiologists. A Study of Scientific Ideas* (Berkeley, Los Angeles and London, 1980)
Hooykaas, 'Het Begrip'	R. Hooykaas, 'Het Begrip Element in zijn historisch-wijsgeerige ontwikkeling', Doctorate diss. University of Utrecht, 1933, Engl. tr. by H.H. Kubbinga, private circulation
HP	The University of Sheffield, The Hartlib Papers. A Complete Text and Image Database
Lasswitz, *Geschichte*:	K. Lasswitz, *Geschichte der Atomistik: vom Mittelalter bis Newton,* 2 vols. (Hamburg, 1890)
Mersenne, *Correspondance*	*La correspondance du P. Marin Mersenne: Religieux Minime* edited by C. de Waard *et alii,* 10 vols. (Paris, 1921-67)
Metzger, *Doctrines Chimiques*:	H. Metzger, *Les doctrines chimiques en France du début du XVIIe à la fin du XVIIIe siècle* (Paris, 1969^2)
Newman, *Gehennical Fire*	W.R. Newman, *Gehennical Fire. The Lives of George Starkey, an American Alchemist in the Age of the Scientific Revolution* (Cambridge, Mass., 1994)
Oldenburg, *Correspondence*	*The Correspondence of Henry Oldenburg*, edited by A.R. Hall and M. Boas Hall, 13 vols. (Madison, Milwaukee and London, 1965-86)
Partington	J.R. Partington, *A History of Chemistry*, 4 vols. (London, 1961-70)
Robert Boyle Reconsidered	M. Hunter (ed.), *Robert Boyle Reconsidered* (Cambridge, 1994)
Webster, *Instauration*:	C. Webster, *The Great Instauration. Science, Medicine and Reform 1626-1660* (London, 1975)
Works	*The Works of the Honourable Robert Boyle,* ed. T. Birch, 6 vols. (London, 1772)

INTRODUCTION

It is a widely accepted view that for much of the seventeenth century chemistry and atomism were dominated by two different theories of matter. The former was dominated by vitalistic ideas, the latter by mechanical theories. According to this interpretation, in the first decades of the seventeenth century chemistry was not yet part of the new science, but was either a purely practical discipline or a confused *mélange* of philosophical and mystical doctrines. On the other hand, the revival of atomism in the early seventeenth century was the beginning of a process leading to the establishment of mechanical philosophy. The mechanical philosophy replaced the qualitative theories of matter of both the Aristotelians and the Paracelsians. Atomism expanded into the mechanical philosophy, which reduced all natural phenomena to matter and motion. It rejected the scholastic notion of substantial forms and explained sensible qualities in terms of motion of corpuscles endowed with purely mechanical properties.[1] Evidently, this interpretation considered Descartes's mechanism as the prototype for understanding the mechanical philosophy which flourished in the second part of the century. Historians have paid attention to Descartes's and Gassendi's different metaphysical views, but, with a few exceptions (as for instance O. Bloch), have failed to evaluate the difference between their theories of matter.[2] This difference is by no means insignificant and its correct evaluation may help us to understand the development of corpuscularianism in the second half of the seventeenth century. While for Gassendi matter is active, for Descartes it is inert. Indeed, the theory of active matter played an important part in seventeenth-century natural philosophy, notably in England.

According to the standard view of the scientific revolution, the key figure in the establishment of mechanical philosophy was Robert Boyle, who used chemical experiments to support the mechanical theory of matter. In turn, the mechanical philosophy was the basis of chemistry, which was thus freed from the obscurities of Paracelsianism and became part of the new natural philosophy. As Richard S. Westfall claimed in 1971, "The story

[1] See for instance M. Boas, 'The Establishment of the Mechanical Philosophy', *Osiris* 10 (1952), 412-541.

[2] O. Bloch, *La Philosophie de Gassendi. Nominalisme, Matérialisme et Métaphysique* (The Hague, 1971).

of chemistry in the second half of the century is the story of its conversion to the mechanical philosophy."[3]

It is my contention that this view, which is still accepted by the majority of historians, oversimplifies the history of early modern atomism and fails to take into account the complex relationships between corpuscular philosophy, Aristotelianism and chemistry. A reassessment of the roles of alchemy and iatrochemistry in the scientific revolution had already started in the previous decades. Following the path of Walter Pagel, scholars of Paracelsianism and of van Helmont's iatrochemistry have stressed the importance of the so-called chemical philosophy in Renaissance and early modern science and medicine, providing detailed studies of the Paracelsian movement in various countries. Nevertheless, the standard view of early seventeenth-century chemistry as opposed to atomism is still reasserted in the influential studies of A.G. Debus.[4] It is true that Paracelsus and the early Paracelsians never adopted corpuscular theories of matter, but a number of Paracelsian doctrines were accepted by early seventeenth-century atomists, such as Hill, Sennert and Basso. In addition, as Graham Rees has shown, Francis Bacon's matter theory encompassed some relevant Paracelsian doctrines.[5] Finally, in the second half of the seventeenth century Paracelsian and corpuscular theories were often amalgamated. The connection between atomism and chemistry (and alchemy) has not escaped the attention of Hooykaas and Meinel, who have focused on the importance of chemistry in the establishment of early modern atomism.[6] More recently, W. Newman has maintained that alchemists and iatrochemists, following the ps-Geber's *Summa Perfectionis*, held a corpuscular theory of matter, which in his view provides the background to Boyle's chemical work.[7] It is my contention that Newman's studies overemphasise the importance of corpuscular views in the alchemical tradition, though undoubtedly they have the great merit of illuminating sources of early modern corpuscular

[3] R.S. Westfall, *The Construction of Modern Science* (Cambridge, 1971), p. 69.

[4] A.G. Debus, *The Chemical Philosophy*, 2 vols. (New York, 1977).

[5] G. Rees, 'Francis Bacon's semi-Paracelsian Cosmology', *Ambix* 22 (1975), 27-39.

[6] R. Hooykaas, 'Het Begrip'; id., 'The experimental origin of chemical atomic and molecular theory before Boyle', *Chymia* 2 (1949), 65-80; C. Meinel, 'Early Seventeenth-Century Atomism. Theory, Epistemology and the Insufficiency of Experiments', *Isis* 79 (1988), 68-103.

[7] W.R. Newman (ed.), *The* Summa Perfectionis *of Pseudo-Geber* (Leiden, 1991) and id., *Gehennical Fire.*

philosophy other than the classical atomism. Explanations of chemical phenomena by means of a vague corpuscular theory can in fact be found in the *Summa Perfectionis*. However, this was an exception. Corpuscular philosophy was not a major component of alchemy before the seventeenth century.

It was with Angelo Sala and Daniel Sennert that chemistry and corpuscular philosophy became associated. When Robert Boyle advocated the fusion of chemistry and corpuscularianism, he was in fact developing a tradition which Sala and Sennert had initiated at the beginning of the century. In Sennert's works (which are investigated in the first chapter of the present book) we find that chemical experiments provide evidence for the existence of atoms, and in turn a number of chemical processes are explained in terms of particles of matter. Sennert's atomism was not mechanical, but qualitative – a direct legacy of the *minima naturalia*. However, the notion of *minima*, which the studies of Maier, van Melsen and Emerton have illuminated, ought not to be interpreted as identical with that of corpuscles.[8] There are at least three distinct versions of the *minima naturalia* theory: 1. substances cannot exist below a certain minimal size; 2. substances are not stable below a certain limit of magnitude; 3. *minima* are physical indivisible components of bodies, keeping their form. Evidently it is the third version (employed by J.C. Scaliger) which paved the way to the corpuscular theory of matter. Following Scaliger, Sennert explained a variety of natural phenomena in terms of addition and subtraction of atoms which have forms and chemical properties.

The study of French atomism and chemistry in the first half of the seventeenth century (see chapter 2) has revealed that in the age of Descartes a variety of versions of atomism flourished in France – most of them having little in common with Descartes's mechanical philosophy. Unlike Descartes's particles of inert matter, the atoms we find in the works of early French atomists have qualities and powers. Two relevant figures of early seventeenth-century French science, Sebastien Basso and Etienne de Clave, adopted a qualitative version of atomism in their works, and combined atomism with chemistry. They were not alone. In the context of a wide diffusion of Paracelsian ideas in France, several little-known physicians and chemists who corresponded with Mersenne (Christophe Villiers, Théodore Deschamps, Jean Brun, Henry Stanihurst), supported the corpuscular theory

[8] See A. Maier, *Die Vorläufer Galileis im 14. Jahrhundert* (Rome, 1949), pp. 155-215; A.G. van Melsen, *From Atomos to Atom* (Pittsburgh, 1952); N.E. Emerton, *The Scientific Reinterpretation of Form* (Ithaca and London, 1984).

of matter. Like Basso and de Clave, their version of atomism was not mechanical but qualitative. This was also the case with another correspondent of Mersenne, Jean Rey, who published the famous *Essays* (1630), dealing with the calcination of metals. The notion of spirit played a central part in their works and the chemical principles were not rejected. From being a semi-divine substance (as posited by Marsilio Ficino), spirit became the principle of life and movement in the natural world. Investigation of the spirit of the world – of its supposed chemical properties and of the appropriate techniques to 'capture' it – was fairly frequent among French chemists, such as d'Espagnet, Nuysement, Rochas and Nicaise Lefebvre.

In the early decades of the seventeenth century the notion of *semina rerum* became part of the corpuscular theory of matter. This view presupposes that matter is not homogeneous as some particles are endowed with powers, particularly with a formative force. The reinterpretation of *semina* in corpuscular terms, a theory which took hold in the early part of the seventeenth century (by Anselmus Boethius de Boodt, Daniel Sennert, Etienne de Clave and others), played a crucial role in seventeenth-century atomism. Many atomists had recourse to the notion of seminal principles in order to explain a variety of phenomena, both in chemistry and in medicine. Evidently atomists like Pierre Gassendi, Walter Charleton and Nathaniel Highmore were aware of the weakness of the purely mechanical theory of matter. One of the main benefits of the concept of *semina* was that it enabled atomists to introduce teleology within the corpuscular theory of generation – as happened with Robert Boyle.

Early English atomism (which is dealt with in chapter 3) shows unambiguously that Paracelsian views (*semina*, spirit and, in general, the vitalistic conception of nature) coexisted with atomic doctrines. This is the case, among others, with Nicholas Hill and Walter Warner. Francis Bacon explicitly stresses the weakness of mechanical atomism and has recourse to the notion of spirit, as well as to the activity of matter. The view of matter as active plays a relevant part in the atomism that flourished in the 1650s, as demonstrated in the works of Margaret Cavendish and Walter Charleton. The origin of motion in matter was of course a crucial problem for such a pious naturalist as Robert Boyle – who criticised the activity of matter as a notion dangerous to the Christian Religion.

Robert Boyle plays a central part in this book. In chapter four I expand the interpretation of Boyle's theory of matter which I put forward in my article on Boyle's chemistry and corpuscular philosophy published in 1990.[9] In it a revision was proposed of the standard view of Boyle as 'orthodox' mechanical philosopher. The article tried to show that Boyle's theory of matter is corpuscular, not strictly mechanical, as it includes agents endowed with formative power, that is, the seminal principles. The conclusions reached in 1990 have been confirmed by further investigations of Boyle's works; hence Boyle's chemistry may be described as corpuscular, rather than mechanical.

Subsequent research has highlighted the importance of the notion of texture in Boyle's theory of matter. Rather than reducing a phenomenon or a quality to a given shape and size of the particles – the approach adopted by Descartes and his followers – Boyle focused on the texture of corpuscles, that is to say, on the different kinds of aggregates of particles. Boyle's chemical and medical works clearly show that he did not subordinate chemistry to the principles of mechanical philosophy. He explained chemical processes in terms of corpuscles endowed with chemical, not just mechanical properties. When Boyle spoke of corpuscles of mercury, sulphur, air, etc., he did not refer to the simple particles of universal homogeneous matter. He meant compound corpuscles, which are made of the primary simple particles. The simple corpuscles have only mechanical properties: shape, size and mobility.[10] The difference between simple and compound corpuscles – which is central for an understanding of Boyle's chemical theories – is based on his classification of corpuscles. This classification has its roots in early seventeenth-century atomism, and in particular, in the works of Sennert and Basso. Boyle's hierarchy of corpuscles is well documented in *The Origine of Formes and Qualities* (1666). By their close union, the simple particles, or corpuscles of the first order form primitive concretions or clusters of particles (that is, corpuscles of the second order). The latter are not easily broken apart, but remain unchanged in the natural bodies. These corpuscles form clusters of a higher order. Unlike the corpuscles of the first order, the corpuscles of the second and of a higher order possess not merely mechanical but also chemical properties.

[9] A. Clericuzio, 'A Redefinition of Boyle's chemistry and corpuscular philosophy', *Annals of Science* 47 (1990), 561-89.

[10] For the notion of mobility in Boyle's theory of matter, see P. Alexander, *Ideas, Qualities and Corpuscles. Locke and Boyle on the External World* (Cambridge, 1985), pp. 68-70.

It is well-known that Boyle opposed the classification adopted by chemists, since it was based on a small number of analogies and overlooked the differences. Boyle's critique of the acid/alkali theory is also well-known. This does not mean that he dismissed the quest for a chemical classification. Though he did not produce a comprehensive classification, he proposed new criteria based on more sophisticated methods of analysis and on newly-discovered chemical indicators.

Boyle's criticism of the chemical principles did not mean that he denied the existence of simple and homogeneous substances. He rejected the chemists' claim that the substances they obtained were simple and homogeneous. He also criticised the chemists' techniques of obtaining them from mixed bodies, but did not peremptorily deny their existence. There is at least one substance which he regarded as simple and homogenous – mercury.

A redefinition of Boyle's corpuscular philosophy and chemistry makes it possible to assess the impact of his ideas in the second half of the seventeenth century and contributes to the understanding of the complex vicissitudes of chemistry at the turn of the century (chapters 5 and 6). In the last decades of the seventeenth century the chemical doctrine of the principles, as well as the Helmontian theories, did not disappear. In England and in Germany the combination of Helmontianism and Boyle's views was fairly widespread. Van Helmont's concepts of seeds, spirit and *Archeus,* as well as the *Alkahest*, were reinterpreted in corpuscular terms – as verified by Daniel Coxe, a hitherto neglected chemist and correspondent of Robert Boyle. After the publication of *The Sceptical Chymist* (1661) the Paracelsian principles were still maintained by a number of English and Continental chemists and physicians – albeit in substantially revised form. This demonstrates that the arguments in *The Sceptical Chymist* had some impact on chemistry. Boyle's criticisms of chemical principles were in fact accepted by many chemists who explicitly denied that salt, sulphur, mercury, water and earth were the ultimate constituents of mixed bodies. When late seventeenth-century chemists (such as Lemery) adopted the principles in their explanations of chemical reactions, the status of principles had changed considerably from what it was in the Paracelsian tradition. The view that they were simple and homogeneous substances was discarded by the majority of chemists who saw them as 'working tools'. This was the case with some influential textbooks which appeared (mainly in France) after the publication of *The Sceptical Chymist*, as well as in a number of chemical works published in Europe in the last part of the century.

Hopefully, three final remarks may minimise any misunderstanding of the views here expressed. 1. Any reference to the strict mechanical philosophy means a theory of matter (like the Cartesian one) according to which matter is inert and all interactions in nature are produced by the impact of particles. 2. The term reductionism does not refer to the current philosophical meaning of this notion. In the present context reductionism means the practice of trying to show that certain entities (chemical qualities, etc.) may be eliminated by reducing all reference to them to some other (more basic) entities – that is, the mechanical properties of matter: shape, size and motion. Hence, Descartes's and the Cartesians' science may be described as reductionist. Of course, generally speaking, not all reductionists are mechanists. However, for the scope of the present study, which is confined to theories of matter in the seventeenth century, it is possible to ignore other forms of reductionism. 3. Chemistry, or chemical works also mean alchemy and alchemical works. As Principe and Newman have maintained, the difference between chemistry and alchemy was blurred in the seventeenth century – though that does not imply that the difference did not exist, nor that it was not perceived at all at the time.[11]

[11] W.R. Newman and L. Principe, 'Alchemy vs. Chemistry: the Etymological Origins of a Historiographic Mistake', *Early Science and Medicine* 3/1 (1998), 32-65.

CHAPTER 1

MINIMA TO ATOMS: SENNERT

INTRODUCTION

The traditional view, which continues to have many influential defenders down to the present day, is that the corpuscular theories which flourished in the early seventeenth century have their roots in the rediscovery of ancient atomism. This view is certainly correct, but it fails to understand the complex origins of early modern atomism. During the last decades, however, there has risen a different approach to seventeenth-century atomism which has focused on the importance of *minima naturalia*. In the paragraphs which follow I shall investigate two distinct doctrines which contributed to the emergence of corpuscular theories of matter in the seventeenth century, namely, *minima naturalia* and *semina rerum*.

The doctrine of *minima* never questioned Aristotelian notion of form and was usually seen as distinct from atomism.[1] Nevertheless, in the second half of the sixteenth century Scaliger interpreted *minima* as particles – an interpretation which opened the way to the corpuscular philosophy which emerged in the next century.

With Daniel Sennert the transformation of *minima* into atoms became unequivocal. Sennert's atomism was different from classical atomism. His atoms are particles of matter, with different qualities which remain unchanged in compound bodies. Atoms of the four elements keep their own forms in the compound. Sennert maintained that 'subsidiary forms' persist in mixed bodies under the 'jurisdiction' of the 'dominant form'. As we shall see, the qualitative version of atomism – which is in fact a direct legacy of the *minima naturalia* theory – was usual in the first half of the seventeenth century and was later adopted by Sir Kenelm Digby and George Starkey.

The Aristotelian notion of *semina rerum* had a complex history as it was adopted in various philosophical contexts. Consequently, the concept of *semina rerum* gradually received different interpretations. After Augustine *semina* were generally conceived as immaterial active principles having nothing in common with atoms; but in medicine, notably in the works of

[1] Cf. the special issue of *Early Science and Medicine*: "The Fate of Hylomorphism", eds. C.H. Lüthy and W.R. Newman, 2/3 (1997).

Fracastoro which dealt with contagion, *semina* received a clear corpuscular interpretation – which in fact originated from Lucretius. The notion of *semina* contributed to the development of atomism in two different ways. First, the focus on units of matter as agents of contagion stimulated the interpretation of chemical and biological phenomena in terms of addition and subtraction of corpuscles. Second, the notion of *semina* contributed to the emergence of the interpretation of atoms as corpuscles endowed with force and formative power – a view that, as we shall see in the ensuing chapters, is fundamental to a large part of seventeenth-century theories of matter.

CHANGING VIEWS OF *MINIMA NATURALIA*

The transformation of *minima naturalia* into physical units marked a significant step towards the establishment of corpuscular philosophy.[2]

The notion of *minimum* is based on the Aristotelian doctrine that substantial forms are not preserved beyond a given limit – both *ad maximum* and *ad minimum*. This applies to compound bodies as well as to *homogenea* (the four elements). The original meaning of the concept of *minimum* is that of limit to the division of a substance.[3] With the exception of a small number of authors, medieval philosophers took *minima* to be not real parts, but the minimal size below which forms could not be maintained. Whereas Scotists such as Walter Burleigh (14th century) adopted the concept of *minimum* only for heterogeneous substances, a significant change in the view of *minimum* was introduced by Buridan and Albert of Saxony, both pointing to an interpretation of *minimum* in physical terms. Buridan claimed that substances are not stable below a given limit of magnitude. Albert of Saxony stated that the *minimum* is not an absolute concept, as it depends on 'environment and conditions', so that a certain quantity of a substance which is too small to exist in one environment could very well be stable in another.[4]

The available evidence does not allow us to conclude that the medieval theory of *minima* can be considered as a corpuscular theory of matter. The

[2] See A. Maier, *Die Vorläufer Galileis im 14. Jahrhundert* (Rome, 1949), pp. 155-215; and A.G. van Melsen, *From Atomos to Atom* (Pittsburgh, 1952).

[3] See van Melsen (n. 2), pp. 59-60 and Maier (n. 2), pp. 183-4.

[4] J. Buridan, *Questiones totius libri Physicorum*, lib. I, q. XIII and Albert of Saxony, *Acutissimae Questiones super libros de physica auscultatione*..., both texts are quoted in van Melsen (n. 2), pp. 62-3.

main reason is that even if *minima* were deemed to be actually existing parts of matter, they were not used as *explanans* of natural phenomena. In addition, as Anneliese Maier has pointed out, the majority of scholastic philosophers did not see the notion of *minimum* as incompatible with the Aristotelian doctrine of the infinite divisibility of natural bodies.[5] An important exception in medieval science is Ps-Geber's *Summa Perfectionis* (written around the end of the 13th century), where a number of chemical operations are explained in terms of what Hooykaas called 'vague corpuscular theory', that is, by means of particles of mercury and sulphur.[6] The view of *minima* as the ultimate units of matter occurs in Giovanni Agostino Panteo's *Ars transmutationis metallicae* (1518). In this work Panteo maintains that *mixtio* is produced "per minima, id est per indivisibilia."[7]

Agostino Nifo and Julius Caesar Scaliger mark two important steps towards the corpuscular interpretation of *minima naturalia.* Nifo maintained that generation, growth and alteration take place by means of *minima.*[8] Nifo's view that *minima* are present as parts paves the way to Scaliger's subsequent solution of the problem of *mixtio.*[9]

Scaliger's version of the *minima naturalia* theory was both innovative and influential. In *the Exotericarum Exercitationum Libri XV* Scaliger

[5] See Maier (n. 2). Though Emerton stresses the difference between medieval *minima* and atoms, she defines the *minima naturalia* doctrine as the Peripatetic corpuscular theory. See N.E. Emerton, *The Scientific Reinterpretation of Form* (Ithaca and London, 1984), pp. 88; 91-2. In my opinion, *minima* (in scholastic philosophy) are not adopted as the foundation of a corpuscular theory.

[6] R. Hooykaas, 'Het Begrip', pp. 37-40 (quote on p. 40). See also W.R. Newman (ed.), *The* Summa Perfectionis *of Pseudo-Geber* (Leiden, 1991) and id., *Gehennical Fire*, pp. 92-106; and H.H. Kubbinga, 'La théorie de la matière de Geber', in Z.W.R.M. van Martels (ed.), *Alchemy Revisited* (Leiden, 1990), pp.133-8.

[7] G.A. Panteo, *Ars et Theoria Transmutationis Metallicae...*, in *Theatrum Chemicum...* (Strasbourg, 1619), ii, pp. 528-630 (hereafter TC), quotation from p. 534. On Panteo, see A. Perifano, *L'Alchimie à la Cour de Côme Ier de Médicis: savoirs, culture et politique* (Paris, 1997), pp. 18-19.

[8] A. Nifo, *Expositio super octo... libros de Physico Auditu* (Venice, 1569), p. 576b, and 476b. Cf. van Melsen (n. 2), pp. 65-9.

[9] Cf. van Melsen (n. 2), pp. 66. Nifo explains a variety of chemical phenomena by referring them to the different degrees of porosity in natural bodies. For Nifo, unlike the atomists, pores are not void, they contain matter, though more rarefied. See Nifo, *In Libris Aristotelis Metereologicis Commentaria* (Venice, 1551), pp. 139b-140a.

maintained that *minima* (of both the elements and the compounds) were not just the limit to division, but were the actual physical indivisible components of bodies.[10] He explained a wide range of physical and chemical phenomena by having recourse to the *minima naturalia.* His solution to the vexed question of the *forma mixti* led to the emergence of corpuscular theories – though Scaliger himself stressed the difference between his own view of the *mixtio* and that of the ancient atomists.[11] According to Scaliger, "mixtion is the motion of the minimum bodies towards mutual contact so that union is achieved." Atoms cannot produce a continuous body.[12] He rejected Avicenna's view of the permanence of elements' forms in the composition, by arguing that the result would be a heap – as in the atomic theory – not a real *mixtum.* He believed that the distinctive property of *mixtio vera* is the continuity of its constituent parts.

Scaliger's version of *minima* contains a relevant innovation, which is the importance he gave to the motion, size and arrangement of *minima.* He believed that *minima* of different substances, including the four elements, differ in size. The *minima* of earth are the biggest, followed by those of water, air and fire.[13] The arrangement of *minima* explains the different states of aggregation of bodies. For instance, rain, hail and snow are all water, the difference, according to Scaliger, being simply in the way their *minima* are joined. Scaliger's explanation is based on the density of the body, which in turn is determined by the greater or lesser distance among the constituent *minima.*[14] Both density and rarity are produced by the motion of *minima*: "Rarefactio motus est, quo rarefacta promovent terminos suos: densatio vero motus, quo densata contrahunt terminos suos."[15]

Scaliger also referred to the porosity of bodies being the cause of some chemical reactions, for example the production of heat when quicklime is

[10] J. C. Scaliger, *Exotericarum Exercitationum Libri XV de Subtilitate* (Paris, 1557), p. 35^{r}.

[11] For medieval discussions of *forma mixti* see A. Maier, *An der Grenze von Scholastik und Naturwissenschaft* (Rome, 1952^{2}), pp. 3-140.

[12] "Mistio est motus corporum minimorum ad mutuum contactum ut fiat unio. Neque enim velut atomi Epicureae sese contingunt: ita corpuscula nostra, sed ut continuum corpus, atque unum fiat." Scaliger (n. 10), p. 143^{v}.

[13] Ibid., p. 33^{v}.

[14] Ibid., p. 356^{r}.

[15] Ibid., pp. 25^{v}-26^{r}.

mixed with water. This occurs because the parts of fire which are contained in the pores of quicklime leave it when water and quicklime are combined.[16]

The concept of *minima* is extensively employed by the French alchemist Gaston Duclo, active in the second half of the sixteenth century.[17] Following Scaliger, Duclo (*Apologia,* 1590) explained *mixtio* in terms of contact of *minima*. Nonetheless, he maintains that the *vera mixtio* requires the emergence of a new form.[18] An important role in Duclo's theory of matter is played by the notion of *semen*. In *Apologia* and in *De recta et vera ratione...* (1592) Duclo stresses the importance of *semina* in the transmutation. I do not share Principe's view that Duclo's use of *semina* is purely metaphorical.[19] Though Duclo rejected the view that metals are alive, he believed that in seed of gold and of silver is contained a specific *vis*.[20]

SEMINA RERUM

The other relevant contribution to the origin of modern corpuscular philosophy was the re-interpretation of *semina* as corpuscles. This interpretation was by no means a linear and straightforward process, since, with the exception of Epicureanism, *semina* were usually seen as distinct from atoms. A wide range of views of *semina* may be found in different historical contexts. Meaning also changed according to the intellectual contexts (medicine, philosophy or theology). Two opposite meanings of the notion of seeds may be found. *Semina* were conceived either as non-material, semi-divine, entities, or as active, living particles of matter. Common to the various interpretations of this notion was the idea that seeds

[16] Ibid., p. 9^r.

[17] On Duclo see L. Principe, 'Diversity in alchemy. The case of Gaston "Claveus" Duclo, a scholastic mercurialist alchemist', in A.G. Debus and M.T. Walton (eds.), *Reading the Book of Nature. The other Side of the Scientific Revolution* (Kirksville, Missouri, 1998), pp. 181-200.

[18] G. Duclo, *Apologia chrysopoeiae...*, TC, iii, p. 33.

[19] Principe (n. 17), p. 196-7. Principe describes Duclo's transmutational theory as "physical, rational, and coherent".

[20] Duclo, *Apologia*, TC, iii, p. 25. In *De recta...*, Duclo affirms that though *semina* are not sufficient for the generation, they are certainly a necessary agent (TC, iv, pp. 448). Duclo does not refrain from adopting biological notions, like those of *concoctio* and *digestio* (ibid., p. 448). It is my contention that the use of vitalistic views was not incompatible with the corpuscular theory, even less in works (like Duclo's), where the corpuscular theory of matter was not fully developed.

were invisible living entities, endowed with some kind of formative power or potentiality of growth.[21]

The use of seeds in Greek cosmology can be traced back to Anaxagoras, who maintained that all natural bodies are generated from specific seeds.[22] In Epicurus and Lucretius the notions of *sperma* and *semen* are equivalent to *atomos*. Epicurean atoms were not inert, but conceived as units of matter endowed with activity.[23] More relevant is Lucretius's use of *semina*. In *De rerum natura*, *semen* was one of the terms Lucretius employed to translate *atomos*. Lucretius also used *semina* to mean aggregates of atoms. He argued that the combinations of atoms with different shapes form different seeds. From *semina* all living bodies have their birth – including humans. Lucretius believed that some *semina* are injurious to men and produce various diseases – a view later adopted by Fracastoro. In his extensive discussions of the origin of diseases, including plague, Lucretius spoke of pestilential seeds putrefying air. Such deadly *semina* – he claimed – originate both from the putrefied soil and from the sky.[24] It is apparent that Lucretius maintained that there are formative powers within the atoms, since for him matter is endowed with activity and powers.

The other notion of *semina* which emerged in Greek thought was the Stoics' theory of *logoi spermatikoi* – a theory which had a stronger impact than the Epicurean and Lucretian views of *semina*. Stoics postulated that matter was a continuum, therefore indefinitely divisible. On the basis of their monistic view of reality (matter and active forces being two aspects of

[21] An excellent study of the concept of *semina* is H. Hirai, 'Le concept de semence dans les théories de la matière à la Renaissance' (unpublished doctoral dissertation, University of Lille, 1999).

[22] H. Diels and W. Kranz, *Die Fragmente der Vorsokratiker* (Zurich and Berlin, 1964[11]), fr. 4 and 21a. Cf. G. Vlastos, 'The Physical Theory of Anaxagoras', *The Philosophical Review* 59 (1950), 31-57; and G.E.R. Lloyd, *Polarity and Analogy. Two Types of Argumentation in Early Greek Thought* (Cambridge, 1966), pp. 244-7.

[23] See Epicurus, *Letter to Herodotus*, in D.L., X, 38 and 74; and A. Alberti, *Sensazione e realtà. Epicuro e Gassendi* (Florence, 1988), pp. 69-72. Lloyd (n. 22) maintains – mainly on the grounds of Aristotle's report – that the notion of seed was already associated with atoms by Leuccipus and Democritos: "Like Anaxagoras, and conceivably under his direct influence, the Atomists appear to have used an image of seeds, or rather of a seed-mixture, in connection with the primary substance, the atoms themselves…" (p. 247).

[24] Lucretius, *De rerum natura*, vi, 655-666, 769-830, 1090-1137. Cf. V. Nutton, 'The seeds of disease: an explanation of contagion and infection from the Greeks to the Renaissance', *Medical History* 27 (1983), 1-34, esp. 9-10.

the same substance), Stoics conceived *logoi spermaticoi* (seminal reasons) as generative forces, principles of activity in matter, which were responsible for the generation and the subsequent development of natural bodies. Stoics' seeds specify the divine creative power which is immanent in nature, as they contain a rational 'programme'.[25]

In Plotinus's and Augustine's philosophy the seminal reasons are dematerialised. Plotinus (who denied the existence of atoms), believed that the seminal reasons originate from the *nous*, which bequeaths them to the world soul. Seminal reasons are therefore conceived as links between the intelligible and the physical worlds. In the generation of living bodies it is the soul which operates in the seeds by transforming and by moulding matter.[26]

Augustine adopted and transformed the notion of seminal reasons to demonstrate that God is the only creator. His idea that *semina* as immaterial beginnings placed by God in matter at the outset was meant to emphasise God's causative power in nature. Natural species were created at the beginning of the world in the form of seeds. Then different bodies emerged from their seminal reasons.[27] Augustine's version of *semina* influenced subsequent discussions of this topic throughout the Middle Ages and inspired the Paracelsians' and van Helmont's notion of seeds.

In scholastic philosophy the notion of seminal reasons was used in embryology (Giles of Rome), in mineralogy (Albertus Magnus) and to

[25] "These *logoi* contain within themselves the germs of everything they are to become. They account for normal, unexceptional growth and development... They also account for exceptional events. In this case, the *logoi spermatikoi* are understood as individual seeds planted by the divine *logos* with a delayed reaction." M.L. Colish, *The Stoic Tradition from Antiquity to the early Middle Ages*, 2 vols (Leiden, 1985), ii, p. 32.

[26] Plotinus, *Ennead* ii. 3, 16-17; iii. 4, 7; iv. 3, 10; v. 9, 6.

[27] "All things that come to corporeal and visible birth have their hidden seeds lying dormant in the corporeal elements of this world. There are of course the seeds plants and animals produce which we can see with our eyes; but of these seeds there are other hidden seeds from which, at the creator's bidding, water produced the first fishes and birds, and earth the first plants and animals of their kind. Nor was this basic seminal force exhausted in that primordial breeding...", Augustine, *De Trinitate*, iii. 2. 13. On Augustine's seminal reasons see C. Boyer, 'La théorie augustinienne des raisons séminales', *Miscellanea Agostiniana*, 2 vols (Rome, 1931), ii, pp. 795-819; J.M. Brady, 'St. Augustine's theory of seminal reasons', *New Scholasticism* 38 (1964), 141-58 and Colish (n. 25), pp. 204-6.

explain the origin of forms (Theodoric of Freiberg).[28] Albertus's theory of *semina* is of special interest as it combines Aristotelian views with alchemical doctrines – which were in turn influenced by stoicism. When in the *Book of Minerals* Albert formulates the theory of mineralizing power as the efficient cause of minerals, he adopts a biological model:

> The mineralizing power is a certain power, common to the production of both stones and metals, and of things intermediate between them. And we say in addition that if this is active in forming stones, it becomes a special power for producing stones. And because we have no special name for this power, we are obliged to explain by analogies what it is. Let us say, then, that just as in an animal's seed, which is a residue from its food, there comes from the seminal vessels a force capable of forming an animal, which actually forms and produces an animal, and is in the seed in the same way as that an artisan is in the artifact that he makes by his art; so in material suitable for stones there is a power that forms and produces stones, and develops the form of this stone or that.[29]

In medieval alchemical texts the 'Philosophers' Stone' was often conceived as the union of masculine and feminine principles. A relevant part of this process was the extraction of the pure seeds of gold and of silver. The seeds were obtained in the form of sulphur and mercury – sulphur being the masculine seed, mercury the female.[30] The analogy between the preparation of the *lapis* and the generation of the embryo is adopted in ps-Lullian *Testamentum.*[31]

[28] See M.A. Hewson, *Giles of Rome and the Medieval Theory of Conception* (London, 1975); P.M. Wengel, *Die Lehre von den rationes seminales bei Albert der Grossen* (Würzburg, 1937); A. Maier, *An der Grenze* (n. 11), pp. 63-7. On Theodoric, see W.A. Wallace, *The Scientific Methodology of Theodoric of Freiberg. A a Case Study of the Relationship between Science and Philosophy* (Fribourg, 1958).

[29] Albertus Magnus, *The Book of Minerals*, translated by D. Wyckoff (Oxford, 1967), p. 22. Cf. J.M. Riddle and J. Mulholland, 'Albert on Stones and Minerals', in J.A. Weisheipl (ed.), *Albertus Magnus and the Sciences. Commemorative Essays* (Toronto, 1980), pp. 203-34.

[30] See E.O. Lippmann, *Entstehung und Ausbreitung der Alchemie,* 3 vols (Berlin, 1919-54), pp. 315-26; J. Read, *Through Alchemy to Chemistry* (London, 1957), p. 19. Hooykaas, 'Het Begrip', p. 49.

[31] Michela Pereira and Barbara Spaggiari, *Il «Testamentum» alchemico attribuito a Raimondo Lullo* (Florence, 1999), pp. 28; 476-8.

The concept of *semina* re-surfaces in Renaissance philosophy and medicine. It was interpreted as an immaterial informing principle, often as an alternative to the Aristotelian notion of form. As the concept of *semina* in the Renaissance has been thoroughly investigated by H. Hirai, in this section I give a brief account of three influential interpretations to this concept, namely those of Ficino, Fracastoro and Paracelsus.[32]

Ficino's concept of *semina,* which was clearly indebted to Plotinus, provides a link between the Neoplatonic (and Augustinian) views and the scholastic doctrine of substantial forms. In his *Theologia Platonica* Ficino maintains that *semina* are hidden in the prime matter, from which they draw the forms of the four elements. For Ficino, *semina* occupy an intermediate position between matter and form.[33]

In Fracastoro's works on contagion – showing strong Lucretian influences – the notion of *semina* is reinterpreted in terms of invisible units of matter. The idea of seeds of disease was not entirely new, as it was already in Lucretius. It was also adopted by Asclepiades of Prusa, by the Methodic school and by Galen – though the latter gave *semina* limited explanatory power. Fracastoro's theory of *semina* as the cause of communicable diseases was both articulate and consistent. Its importance lies in the fact that it was an explanation of contagion which paved the way to the ontological theory of disease as an alternative to the humoral.[34] Fracastoro maintained that seeds were invisible particles which could propagate from a great distance. Some of them (notably *semina* of syphilis) are produced in the sky and propagate due to favourable atmospheric conditions. Once they have penetrated their host, *semina* of diseases can multiply, bringing about the putrefaction of humours.[35]

[32] See Hirai (n. 21), pp. 23-200.

[33] M. Ficino, *Theologia Platonica*, iv, 1.

[34] W. Pagel defines the ontological theory of disease as follows: "Diseases are regarded as entities in themselves distinguishable by specific changes and causes. In this view the main tenet of humoral pathology – that the sick individual determines the nature of disease – is completely reversed: it is now the individual disease that conditions the patient and manifests itself in a characteristic picture." Pagel, *Paracelsus. An Introduction to Philosophical Medicine in the Era of the Renaissance* (Basle, 1982^2), p. 137.

[35] G. Fracastoro, *De sympathia et antipathia rerum, liber unus. De contagione et contagiosis morbis et eorum curatione, libri III* (Venice, 1546). Cf. Nutton (n. 24); and id., 'The Reception of Fracastoro's Theory of Contagion. The Seed That Fell among Thorns?', *Osiris*, 2nd series, 6 (1990), 196-234. Fracastoro's adoption of Lucretian atomism is also attested in his poem *Syphilis, sive morbus gallicus*

The notion of *semina* was central to the works of Paracelsus and of the Paracelsians. Paracelsus's views of seeds are rather complex, as he gives no unequivocal definition of this concept, but employs it in a variety of contexts – chemistry, pathology, embryology and mineralogy. As part of his rejection of Aristotelian elements and qualities, Paracelsus placed special emphasis on *semina*, which he considered as invisible spiritual forces and as archetypes.[36] For Paracelsus, *semina*, which originate in the Word, are contained in the *Yliaster* and are prior to chemical principles and to elements. Nature as a whole is a *panspermia*.[37] Pagel gives a perspicacious interpretation of Paracelsus's *semina*:

> Generalising the hylozoistic semen-principle of the alchemists, Paracelsus poses invisible 'semina' as the germ cells of every object in nature opposing these to the visible elements of the ancients. Instead of units of matter, Paracelsus searches for the 'Logoi' in matter and finds them in the 'Semina' and the 'Intelligences' which they carry.[38]In *Das Buch de Mineralibus* Paracelsus explains the birth of minerals from mineral seeds which are contained in water.[39]

Paracelsus's view of *semina* was developed by Peder Soerensen (Severinus), whose *Idea Medicinae* (1571) contains a detailed and influential treatment of this theory. Severinus conceived *semina* as spiritual entities, not as units of matter. Therefore, he claimed, *semina* are not to be confused with visible seeds. In Soerensen's Neoplatonic interpretation of Paracelsus *semina* are described as the link between the visible world and invisible substances.[40] The qualities of natural bodies do not originate from

(Verona, 1530), where he uses two key Lucretian terms, i.e. *semina* and *primordia rerum*.

[36] Cf. M.L. Bianchi, 'The visible and the invisible. From alchemy to Paracelsus', in P. Rattansi and A. Clericuzio (eds.), *Alchemy and Chemistry in the 16th and 17th Centuries* (Dordrecht, 1994), pp. 17-50.

[37] Paracelsus, *Das Buch de Mineralibus*, in K. Sudhoff (ed.), *Sämtliche Werke* (Munich and Berlin, 1922-3), I, iii, pp. 41-2.

[38] Pagel (n. 34), p. 85.

[39] Paracelsus (n. 37), I, iii, p. 35. On Paracelsus's use of the theory of *semina* in mineralogy see H. Hirai (n. 21), pp. 139-144. Hirai convincingly stresses the analogy between Paracelsus's and Augustine's theories of *semina*. See also D. Oldroyd, 'Some Neoplatonic and Stoic Influences on Mineralogy in the Sixteenth and Seventeenth Centuries', *Ambix* 21 (1974), 128-56.

[40] "Semina sunt vincula utriusque naturae, visibilia invisibilibus coniungentia: in quibus motuum leges, temporum praedestinationes, generationum et

the elements or from "corporum mutua appositione", but from a spiritual agent endowed with "scientia infallibilis".[41]

Following Severinus, a number of Paracelsians gave *semina* a central role both in medicine and in cosmology. The view that diseases generate from specific seeds and that seeds (containing spirits) are responsible for the generation of natural bodies – including metals – became common among Paracelsians and chemical writers at the turn of the seventeenth century.

In typical Paracelsian fashion, Duchesne maintains that seeds containing an architectonic principle go through a continuous circulation, from the heaven to the earth and back to the stars.[42] In J.B. van Helmont's philosophy and medicine *semina* have a central role. Water and *semina* are the two principles of natural bodies, the former being the material one, while seeds are the spiritual principles. Like Severinus, van Helmont's notion of *semina* had unambiguous anti-materialistic character. Van Helmont's rejection of the chemical theory of principles was ultimately based on the view that it bestowed causative power onto material agents.[43] On analogous grounds he attacked the humoral theory of disease. Diseases, he claimed, are generated from specific disease-seeds, namely spiritual and active entities.[44]

Whereas Severinus and van Helmont stressed the spiritual character of *semina rerum* in opposition to the Aristotelian elements, in Michael Sendivogius seeds are deemed to be material agents. He calculated the ratio

transplantationum lithurgiae, et universae mundanae anatomiae dispensationes continentur." P. Severinus, *Idea Medicinae Philosophicae* (Basle, 1571), p. 58. Severinus unambiguously rejected the atomistic theory of matter as materialistic, see ibid., p. 81.

[41] Ibid., pp. 132-6. For Severinus's view of *semina* see W. Pagel, *Harvey's Biological Ideas* (Basle and New York, 1966), pp. 241-44. Pagel convincingly stresses the anti-materialistic orientation of Severinus's theory of *semina*. On Severinus's seeds see also J. Shackelford, 'Seed with a mechanical purpose. Severinus' Semina and Seventeenth-Century Matter Theory', in A.G. Debus and M.T. Walton (n. 17), pp. 15-44.

[42] For Joseph Duchesne (Quercetanus), see A.G. Debus, *The French Paracelsians. The Chemical Challenge to Medical and Scientific Tradition in Early Modern France* (Cambridge, 1991), pp. 51-9 and Hirai (n. 21), pp. 201-22.

[43] J. B. van Helmont, 'Causae et initia naturalium', §§ 20-24, *Ortus Medicinae* (Amsterdam, 1648), pp. 35-6 [hereafter *Ortus*].

[44] van Helmont, 'Ignotus hospes morbus', §§ 62-3 and 66, *Ortus*, pp. 497-9. Cf. W. Pagel, *Jan Baptista van Helmont. Reformer of Science and Medicine* (Cambridge, 1982), pp. 141-3 and G. Giglioni, *Immaginazione e Malattia. Saggio su Jan Baptiste van Helmont* (Milan, 2000).

between each natural body and its seed (1 to 8200).[45] For Sendivogius, *semina* – from which the four elements have their origin – are alive and are endowed with spirit. Metals originate from seeds and the alchemist's task is to extract the metallic seed.[46]

In the late sixteenth century several natural philosophers and chemists who rejected the Aristotelian explanations of the origin of metals and minerals – based on the double exhalation – often adopted the notion of *semina*.[47] In Anselmus Boethius de Boodt's *Gemmarum et Lapidum Historia* (1609) the Aristotelian elements are not entirely rejected (as in Severinus and van Helmont), but are accepted as remote causes. From them are formed the three chemical principles.[48] For de Boodt, the *forma mixti* (and notably the form of stones and minerals) comes from the seminal principles, which contain a formative spirit in them.[49] In de Boodt's *Historia* the corpuscular theory of matter is employed to explain a number of properties of minerals and stones. Gems, he claims, are translucent because they are formed by a perfect union of the *minima* of their constituents. As a result of this union, no pores are left among their particles.[50] Moreover, the different size and motion of the constituent particles are responsible for the different shapes of the gems.[51]

EARLY CHEMICAL ATOMISM: LIBAVIUS AND SALA

Before 1600 embryonic corpuscular views may be found in some alchemical texts, notably in ps-Geber's *Summa Perfectionis*.[52] However, the diffusion of Paracelsianism, with its strong anti-materialistic overtones, did

[45] M. Sendivogius, *De Lapide Philosophorum* (Prague, 1604), repr. in TC, iv, pp. 417-8.

[46] Ibid.

[47] This was true, among others, in the case of Bernard Palissy. See B. Palissy, *Discours Admirable* (Paris, 1580), pp. 122; 134.

[48] A.B. de Boodt, *Gemmarum et Lapidum Historia* (Hannover, 1609), p. 10. On de Boodt see R. Halleux, 'L'oeuvre minéralogique d'Anselme Boèce de Boodt', *Histoire et Nature* 14 (1979), 63-78.

[49] De Boodt (n. 48), pp. 11-2.

[50] Ibid., p. 9.

[51] Ibid., p. 16.

[52] Cf. Hooykaas, 'Het Begrip', pp. 32-40; Newman (ed.), *The* Summa Perfectionis (n. 6); and id., *Gehennical Fire*, pp. 92-106.

not stimulate the development of the chemical corpuscularianism.[53] One has to wait for Andreas Libavius and Angelo Sala to see the adoption of the corpuscular theory in chemistry.

While Libavius's contribution to atomism was certainly marginal, Sala's combination of atomism with chemistry gave substantial impetus to the development of chemical atomism.

Along with Duchesne, Libavius defended (against Jean Riolan the Elder) the interpretation of Democritus as a chemist, and (unlike Duchesne) maintained that Democritean atomism is both confirmed by experiments and compatible to Christian religion.[54]

Libavius's main arguments in support of atomism occur in his *Alchymia triumphans*, where he states that all natural bodies are made of atoms and can in turn be resolved into atoms. His defence of atomism against Riolan does not entail the refutation of the doctrine of the four elements, which he conceived as composed of atoms.[55] It is apparent that Libavius's atomism had much in common with the scholastic theory of *minima*. Despite his effort to vindicate atomism, Libavius made little or no use of atomism as an explanatory theory.

Angelo Sala stands out both for his contributions to practical chemistry and for his adoption of the corpuscular theory of matter as an *explanans* of chemical phenomena. Though he did not reject the notion of form, Sala explained a number of chemical reactions in corpuscular terms. His explanations of the three following chemical processes clearly testify to Sala's commitment to the corpuscular theory of matter: 1. the recovery of

[53] It is worth noticing that Iacob Schengk, who along with Erastus attacked Paracelsus, maintained that the mixed body was formed from *minimae partes*. For Schengk, when the body is divided *in minutissimas partes*, substantial forms do not disappear. See *Tractationum Physicarum et Medicarum tomus unus*... (Frankfurt, 1585), pp. 9; 80-1.

[54] Joseph Duchesne (Quercetanus), *De Priscorum Philosophorum verae medicinae materia*... (Geneva, 1603), pp. 4-5. Quercetanus quoted Suida's *Lexicon* as his source. Though Quercetanus established a link between alchemy/chemistry and Democritus's philosophy, atomism played no part in his works. See also Jean Riolan (the elder) *Ad Libavii Maniam*...(Paris, 1606), pp. 13-17. On Libavius see O. Hannaway, *The Chemist and the Word* (Baltimore and London, 1975); and B.T. Moran, 'Medicine, Alchemy, and the Control of Language: Andreas Libavius versus the Neoparacelsians", in O.P. Grell (ed.), *Paracelsus: the Man and his Reputation* (Leiden, 1998), pp 135-49.

[55] A. Libavius, *Alchymia triumphans* (Frankfurt, 1607), pp. 150-161.

gold dissolved in *aqua regia*; 2. the synthesis and analysis of vitriol; 3. fermentation.

1. Sala's *reductio ad pristinum statum* is based on the Aristotelian distinction between the true homogenous compound, produced by substantial changes – which Sala called transmutations – and the simple juxtaposition of parts. Sala's example of *reductio* is when gold is dissolved in *aqua regia*. The particles of gold disappear, but remain unchanged and can be recovered, for instance, by means of a piece of silver. Sala explained the dissolution of the metal in the acid in corpuscular terms: the acid resolves the metal into small particles, which unite to the reagent. The recovery of gold is described as a transfer of particles.[56]

2. Sala defines vitriol (copper sulphate) as a mixture, that is, an apposition of particles, not a uniform substance (as, he thought, was rock salt). Vitriol consists of spirit of sulphur, water, and copper or iron. Sala was able to synthesise vitriol and in turn to decompose vitriol into its constituent parts. He found by analysis the same ingredients in the same quantity. His explanation was in terms of the apposition of unchangeable particles.[57]

3. According to Sala, fermentation is produced by the motion of particles of bodies. This internal motion of corpuscles, which is generated by heat in the presence of moisture, brings about new arrangements of the particles in a body and the generation of a different and nobler substance.[58] He also

[56] "Reductio autem est operatio quaedam per quam recolligimus & in unam massam coadunamus rem quampiam quae in minutissimas particulas dispersa & dilatata erat, nec non diffusa et mixta alii cuipiam rei, tamquam amissa & perdita fuisset, & interim tamen per Reductionem in pristinum suum statum & essentiam revocatur, & reducitur. Veluti videre in Auro: Hoc enim dissolutio in Aqua Regali (aut Regulo reducto; in liquorem clarum & diaphanum usque adeo ut nihil in eo amplius metallici aut duri appareat, praeter colorem quedam flavum instar Croci) si in id frustum aliquod Argenti injiciatur, confestim Aurum exibit ex isto liquore, & adhaerebit affigetque, se dicto Argento, relinquetque aquam albam & claram: quo facto si quis sumptis hisce duobus ita conjunctis metallis, separet ea a sese mutuo per artem chymicam, inveniemus Aurum istud in pristinam suam formam reductum, ac prorsus tale quale antea exiterat." Sala, *Anatomia Vitrioli* (Leiden, 1617), p. 399. "Aqua Regis aurum [...] atomos solvendo uniat", id., p. 409. On Sala's chemistry, see Hooykaas, 'Het Begrip', pp. 142-8 and Z.G. Gelman, 'Angelo Sala, an iatrochemist of the late Renaissance', *Ambix* 41 (1994), 142-60.

[57] Sala, *Anatomia* (n. 56), pp. 70-3.

[58] "Fermentatio igitur est motus quidam, seu alteratio, a calore interno, in humido agente inducta, qua diversae & inter se pugnantes, substantiae elementares, partim separantur, partim in unum nobiliorem mixtionis modum, ac unionem rediguntur,

noticed that when sugar is fermented and then distilled, a 'phlogistic' spirit is formed.

Sala did not develop an articulate corpuscular theory of matter. He rather employed it to understand chemical reactions. His view of chemical corpuscles was to play a central part in the development of seventeenth-century chemistry.[59]

DANIEL SENNERT

The role of Sennert in the establishment of corpuscular theories of matter has already been recognized by a number of historians.[60] In Sennert's works we find a compromise of Aristotelianism with atomism, and extensive use of chemistry to prove the existence of atoms – both influential in the development of atomism in the first half of the seventeenth century. Sennert's compromise of atomism and Aristotelianism was based on the view that, besides the *forma mixti*, there are subordinated forms. He distinguished two kinds of substantial forms, i.e., subordinate and supervening form – the latter being the principle of organisation.

quod rerum fermentantium strepitu, pugna, & humidi turgescentia apparet, hac mediante res, ad subtiliores, spirituosas, & balsamicas, varieque operandi, & penetrandi virtutes exaltantur", Sala, *Hydrelaeologia*, in *Opera Medico-Chymica*, (Frankfurt, 1647), p. 95.

[59] C. Meinel, 'Early seventeenth-Century Atomism. Theory, Epistemology and the Insufficiency of Experiments', *Isis* 79 (1988), p. 91, maintains that Sala used chemical experiments to confirm the atomistic theory. In my view, what Sala did is exactly the opposite, namely, he explained chemical reactions in terms of transfer of unchangeable particles of matter.

[60] On Sennert see Lasswitz, *Geschichte*, i, pp. 436-54; van Melsen (n. 2), pp. 81-9; T. Gregory, 'Studi sull'Atomismo del Seicento. II. David van Goorle e Daniel Sennert', *Giornale Critico della Filosofia Italiana* 45 (1966), 45-63, esp. 51-63; A.G. Debus, *The Chemical Philosophy* 2 vols. (New York, 1977), i, pp. 191-200; S. Wollgast, *Philosophie in Deutschland zwischen Reformation und Aufklärung 1550-1650* (Berlin, 1988), pp. 438-45; Meinel (n. 59); W.U. Eckart, 'Antiparacelsismus, okkulte Qualitäten und medizinisch-wissenschaftliches Erkennen im Werk Daniel Sennerts (1572-1637)', in A. Buck (ed.), *Die Occulten Wissenschaften in der Renaissance* (Wiesbaden, 1992), pp. 139-57; W.R. Newman, 'The Alchemical Sources of Robert Boyle's Corpuscular Philosophy', *Annals of Science* 53 (1996), 567-85, esp. 575-6. E. Michael, 'Daniel Sennert on matter and form: at the juncture of the old and the new', *Early Science and Medicine* 2/3 (1997), 272-99.

In *Epitome Naturalis Scientiae* (1600) – being a collection of 26 theses defended at Wittenberg in 1599 and 1600 – the theory of *minima naturalia* was used to explain the origin of compounds from the elements.[61] In *Disputatio* XIV, in addition to the statement that, in the mixed body, elements are preserved *in potentia*, we read that "in mistione misciblia primum in parvas & exiguas partes dividi debent."[62] His assertion that forms multiply themselves (*sui multiplicativae*) – which occurs in all Sennert's subsequent works – is in the third dissertation (*De principiis rerum naturalium)*, defended by Iohannes Adam in 1599 (thesis 40). Sennert's view is opposed to that of forms as educed from the potentiality of matter ("formae educuntur e potentia materiae"), and also contrary to the origin of forms from the *dator formarum* or from the heavens – a view held by Jean Fernel. As we shall see, Sennert's view of the origin of forms will develop into the theory of seminal principles in his subsequent works. In *Institutionum Medicinae Libri V* (1611) Sennert dealt with two topics which were to play an important part in his subsequent works: 1. the definition of the role and status of chemistry; 2. the theory of generation. Moreover, Sennert's *Institutiones* bear witness to his early adoption of particles of matter to explain chemical reactions.

Following Libavius, Sennert distanced himself from Paracelsian cosmology and medicine. He recognised that chemistry can implement natural philosophy, but (in opposition to the Paracelsians) he firmly opposed the view that chemistry provides the foundation of medicine and of philosophy.[63] Chemistry, he contended, has two specific ends: the preparation of medicines and the transmutation of metals.[64]

Sennert's view of generation – which was to become a matter of polemics with Freitag – is based on the theory that seed is animated. He believed that *semina* are not generated from the elements, they are made fertile by the *calidum innatum* (which he identifies with the *spiritus insitus*). Having denied that the *spiritus insitus* comes from the elements, Sennert shared the view of *spiritus* as a super-elemental substance. However, he maintained that the soul, not the spirit, is responsible for generation. Spirit is only instrumental. The role of spirit, as Sennert saw it in *Institutiones Medicinae*, was by no means marginal. Spirits acount for a number of

[61] Sennert, *Epitome Naturalis Scientiae*...(Wittenberg, 1600).

[62] Ibid., Disp. XIV, defended by Johannes Schlezerus, on 21 December 1599, see also theses xxii and xxiv.

[63] Sennert, *Institutionum Medicinae Libri V* (Wittenberg, 1611), p. 1032.

[64] Ibid., pp. 1032-3.

phenomena which cannot be explained by means of the elements, like the occult qualities and contagion. According to Sennert, *semina* are animated and propagate *per traducem*.[65]

Following pseudo-Geber's *Summa Perfectionis*, Sennert explains calcination, sublimations and other chemical processes in terms of addition or subctratiion of minimal parts. As Newman has pointed out, Sennert's early corpuscularianism was 'operational' – as the nature and properties of the particles were not defined.[66] Sennert's view of *mixtio*, as contained in the *Epitome Scientiae Naturalis* of 1618, is close to the Averroists'. He maintains that not only the qualities, but also the forms of elements remain in the compound body, though forms are *refractae* – a view which Sennert will modify in subsequent works. Having denied that physical bodies can be divided *in infinitum,* he resorts to the notion of *minima* to account for the origin of mixed bodies, without adopting Scaliger's quantitative interpretation of the *minima* theory.[67] Sennert's view of elements evolves into a compromise between the Aristotelian and the chemical theories. He considers the three chemical principles as the proximate matter of metals, the remote one being the elements.[68] In addition, his view that form is *sui multiplicativa* (as opposed to the doctrine of the generation of forms from the potentiality of matter) presupposes his adoption of the seminal principles as agent endowed with formative power. Forms propagate by means of *semina*. Stones, minerals and metals are generated by an architectonic spirit contained in the seed.[69] It is apparent that, unlike those of

[65] Ibid., pp. 35, 75-79. Sennert's view of *spiritus* is evidently indebted to Marsilio Ficino, *De Vita* (Florence, 1489) and Jean Fernel. See Fernel, *De Abditis Rerum Causis* (Paris, 1548). On Fernel, see D.P. Walker, 'The Astral Body in Renaissance Medicine', *Journal of the Warburg and Courtauld Institutes* 21 (1958), 19-33 and L.A. Deer, 'Academic Theories of Generation: the Contemporaries and Successors of Jean Fernel (1497-1558)', (unpublished Ph. D. dissertation, University of London, The Warburg Institute, 1980); Michael (n. 60), pp. 295-6, maintains that the transmission of the soul *per traducem* had become a Lutheran dogma in the late sixteenth century.

[66] W.R. Newman, 'The Alchemical Sources' (n. 60), 575-6.

[67] "In mistione miscibilia primum in parvas & exiguas partes dividi debent, ideo etiam liquida, fragilia, subtilia, facilius miscentur. Facta sic in portiones exiguas, pro misti natura, Elementorum divisione, eadem per contrarias Qualitates agunt & patiuntur mutuo, se invicem calefaciunt." *Epitome Naturalis Scientiae* (Wittenberg, 1618), p. 225.

[68] Ibid., pp. 345; 379-380; 399.

[69] Ibid., pp. 381-2, 399.

Petrus Severinus, Sennert's *semina* are subsidiary, not alternative, to forms.[70]

In *De Chymicorum* (1619) we find a substantial development of Sennert's thought in two directions.[71] First, Sennert proposes the integration of chemistry within his own eclectic version of the Aristotelian philosophy; second, he gives clear corpuscular interpretation to the notion of *minima naturalia*. Following Libavius, Sennert attacks Paracelsianism, but does not reject chemistry, his explicit goal being to purify chemistry from the Paracelsians' mistakes, notably their cosmological doctrines. Religious preoccupations are behind Sennert's objections to the Paracelsians. The main targets of his polemics are Valentin Weigel, Heinrich Khunrath and Oswald Croll, whom he regards as fanatics, as heretics and as associates of the seditious Anabaptists and the Rosicrucians. In particular, Sennert singles out Croll's doctrine of illumination as the source of knowledge as both erroneous and impious.[72] For Sennert, after the Fall, divine illumination was no longer the source of human knowledge.[73]

The other issue discussed by Sennert in his censure of Paracelsianism was the status of chemistry. For him, chemistry is an art, not a science, and accordingly it does not deal with the principles of natural philosophy and medicine. It can only provide experiments and observations from which philosophers and physicians may derive their own theories. The subject matter of chemistry is twofold: the extraction of essences from natural bodies to be used by physicians, and the transmutation of metals.[74] What Sennert rules out is the Paracelsians' claim that chemistry provides the foundations of medicine and of natural philosophy and that it is not reconcilable with Aristotelianism and Galenism. His detailed historical survey of the origin and development of chemistry is in fact aimed at demonstrating that from antiquity it was a practical discipline and that Paracelsus's chemical teachings were by no means original. He maintains

[70] On Sennert's view of *semina*, see Pagel, *Paracelsus* (n. 34), pp. 333-343.

[71] Cf. W.U. Eckart, 'Grundlagen des Medizinisch-Wissenschaftlichen Erkennens bei Daniel Sennert...', (unpublished Dissertation, Münster, 1977).

[72] On Weigel, see A. Koyré, *Mystiques, spirituels, alchimistes du XVIe siècle allemand* (Paris, 1971). For Khunrath, see DSB. For Croll, see Hannaway (n. 54).

[73] Sennert, *De Chymicorum* (Wittenberg, 1619), pp. 117-121.

[74] Ibid., pp. 14-34. On the status of alchemy and chemistry see J-M. Mandosio, 'L'Alchimie dans la classification des sciences et des arts à la Renaissance', in J-C. Margolin and S. Matton (eds.), *Alchimie et Philosophie à la Renaissance* (Paris, 1993), pp. 11-41.

that chemists who lived before Paracelsus did not question the foundations of Aristotelian philosophy.

Despite his attacks on Paracelsian philosophy, Sennert adopts the three chemical principles, though he uses them in a reduced capacity. Salt, sulphur and mercury, which are formed from the four elements, can explain phenomena that the four elements (and the elementary qualities) cannot explain, i.e. odours, tastes, colours, solidity and inflammability, as well as the properties of several medicines. In the mixed bodies are contained both the elements and the principles. The three principles are subordinate to the elements: they are *prima mixta*, which can be decomposed into the four elements. In natural bodies there is a hierarchy of forms. Therefore, he contends, the forms of the *tria prima* are independent of (though subordinated to) those of the elements.[75] His adoption of the three principles does not prevent him from expressing criticisms of the *tria prima* doctrine. He maintains that salts extracted from plants and animals are neither simple nor homogeneous bodies. In addition, he rejects the view that the three principles, which are obtained by fire analysis, were pre-existent. Sennert also questioned the view of mercury as principle. Mercury, he says, is the proximate matter of metals, but is neither simple, nor a homogenous substance.[76] Sennert replaces mercury with spirit as one of the *tria prima*. He articulates the definition of *spiritus* along the following lines: it is "corpus permeabile, penetrabile, aethereum, purissimum, vivificum, & formae proximum instrumentum."[77] Though his notion of spirit does not replace substantial forms, *spiritus* has a special status in natural philosophy. It is conceived as a semi-divine and active substance, and as the agent directing elements and principles in the generations of mixed bodies. Sennert's adoption of spirit in the theory of *mixtio* is the consequence of his view that mixed bodies do not originate from the forms of the elements. He therefore resorted to something superior to them, that is, spirit.[78]

[75] Sennert, *De Chymicorum*, (n. 73), pp. 264-299.

[76] Ibid., p. 316

[77] Ibid., p. 317.

[78] Ibid., p. 358: "Materiam dant elementa, non vero formam. Mistum, qua mistum, formis elementaribus informari, non repugno. Verum cujusque rei formam specificam, quae rei dat suam essentiam & nomen, ab elementis provenire, nego. Est enim in unaquaque re naturali & partibus corporis praeter materiam, quam elementa suppeditant, divinius quoddam principium & Natura quinta." See also ibid., p. 350-1.

De Chymicorum marks Sennert's explicit acceptance of a corpuscular theory of matter. This is presented as a development of Scaliger's definition of *mixtio*, and is confirmed by a variety of experimental proofs. Indeed, Sennert goes beyond the sixteenth-century version of *minima naturalia* and explicitly advocates the atomists' view that natural bodies are made of unchangeable atoms – though he does not accept the existence of vacuum.[79]

To Sennert the particulate structure of bodies is confirmed by chemical operations, such as sublimation and distillation. Vapours of different bodies, he maintains, retain their nature after being cooled.[80] In mineral waters stones are contained in invisible particles, which if they come together form a hard stony body.[81] As Meinel has shown, the *reductio ad pristinum statum* is one of Sennert's empirical arguments to prove the existence of atoms. One instance of Sennert's use of the *reductio ad pristinum statum* is the recovery of gold and silver from a homogenous alloy by *aqua fortis*. The latter dissolves silver, which can no longer be seen, while gold remains as a powder. Silver too may be precipitated as fine powder from the solution. Sennert obtains gold and silver again from the powders. Hence, he concludes, atoms of gold and of silver retain their own form in the alloy. The process – he claims – is just a case of re-arrangement of unchangeable particles of gold and silver.[82] As we have seen, *De Chymicorum* bears evidence of Sennert's adoption of the atomic theory of matter. Unlike Democritus's atoms, Sennert's are endowed with forms and qualities. There are atoms of the four elements, as well as of the three principles.

In the 1633 edition of the *Epitome Naturalis Scientiae*, Sennert reassesses the question of the forms of elements in the *mixtum*. Whereas in the first edition (1618) he stated that forms remain in the mixed body, but

[79] Ibid., p. 361.

[80] Ibid., p. 362.

[81] Ibid., p. 362: "Procul dubio in talibus aquis mineralis & lapidea materia in minimas particulas resoluta fuit, quae postea suo concursu & synkriseis saxeum & durum corpus constituunt."

[82] Ibid., p. 362: "Si aurum & argentum simul liquescant, ita per minima miscentur, ut visu deprehendi aurum in argento nullo modo possit: si vero postea aqua fortis affundatur, ita solvitur argentum, ut ullum metallum in ea aqua deprehendi visu non possit: cum tamen revera insit & hinc segregatum emergat; & quidem ita, ut aurum & argentum suam naturam retineat; & hoc modo in subtilissimam calcem, quae nihil aliud est, quam congeries aliqua innumerabilium atomorum, redigatur, quae in aurum & argentum purissimum fusione iterum reducitur." The same experiment occurs in Sennert, *Hypomnemata Physica* (Frankfurt, 1636), p. 119. This later version is discussed in Meinel (n. 59), 92-3.

are *refractae*, in 1633 he adopts the Avicennian position, i.e. that forms of elements remain "perfectae et integrae".[83] Sennert explicitly maintains that the ultimate particles are not further divisible. Atoms of the elements remain unchanged in the mixed body.

In *Hypomnemata Physica* (1636) atomism is presented as a comprehensive theory of matter which explains all sorts of phenomena, chemical operations, rarefaction and condensation, as well as spontaneous generation. Forms however do not disappear from Sennert's natural philosophy. He firmly opposes Basso's rejection of substantial forms.[84] In the *Hypomnemata Physica* (as in *De Chymicorum*) Sennert resorts to chemical experiments, that is, to reductions to the pristine state, to support the corpuscular theory of matter. The recovery of metals from solutions in acids is explained in terms of changing arrangements of unchangeable corpuscles. Sennert shows that quicksilver keeps its nature in all changes, namely, as precipitate, sublimate, or as an oil, and can easily be recovered.[85] Besides using chemical arguments to support atomism, Sennert explains by means of corpuscles a wide range of phenomena, like magnetism, poisons and contagion.[86]

A feature of Sennert's corpuscular philosophy found in the *Hypomnemata* is the hierarchy of corpuscles – which is only adumbrated by Sennert but will be fully developed by Robert Boyle. According to Sennert, there are corpuscles of the four elements which are simple and indivisible;

[83] Sennert, *Opera* (Lyons, 1650), i, p. 17. Cf. Michael (n. 60), 289-90.

[84] Sennert, *Hypomnemata* (n. 82), pp. 218-9. For Basso, see below, chapter 2, pp. 39-42.

[85] According to Sennert, metals "redeunt autem & revertuntur in pristinam naturam sublato illo corpore, seu sale, quod a solvendo adhaesit, quae operatio reductio appellatur", because their particles remained unchanged in the solution. The recovery of the original ingredients is possible because the atoms of the same metal tend to unite each other: "Ita si Mercurius sublimatus calci vivae misceatur, & Retortae indatur, sal vitrioli & communis, qui sublimato inest, calci vivae adhaeret, atque ita argentum vivum in pristinam naturam redit & vivificatur; quomodo etiam cinnabaris in argentum vivum reducitur. Calces metallorum in metalla abeunt fusa, dum nimirum igne forti metallo admistum dissipatur, quod tamen facilius sit, si pulvis aliquis fusorius addatur. Sales enim, e quibus illi pulveres fusiori constant, salem, qui calcinato metallo adhaeret, ad se trahunt; a quo liberatae metallorum atomi ob similitudinem uniuntur, & ita in pristinum corpus abeunt." Some substances, for instance mercury, can miss their sensible properties, but can be recovered because they keep their substantial form. Ibid., pp. 109-11.

[86] Ibid., pp. 78-9.

then there are complex and divisible corpuscles, like those of the three chemical principles (called by Sennert *prima mixta*); and finally compound corpuscles of different substances as well as of living bodies. The existence of compound corpuscles is confirmed by purgatives, which pass into a child with the mother's milk. This is explained by Sennert in terms of corpuscles which remain unchanged.[87] It is clear that Sennert's atoms are both chemically homogeneous particles and compound substances.

A large portion of the *Hypomnemata* deals with the problem of generation, one of the most controversial topics of Sennert's philosophy of nature.[88] The seed is composed of spiritual and of material part, and is, according to him, animated by the soul. It is the vehicle of the soul, the latter being the form of a living bodiy. Like forms, the soul is *sui multiplicativa*. As Pagel explained it, souls multiply themselves like images reflected in mirrors.[89] The soul, which is transmitted by the parents, is the formative power operating on matter, employing the spirit as its agent. Sennert clarifies the relationship of *spiritus* (and of *calidum innatum*) to form along the following lines: spirit (a semi-material substance) is subordinate and instrumental to the immaterial formative principle, that is, form, or the soul, in both plants and animals.[90] Sennert believes that spontaneous generation (a mere transformation of matter due to putrefaction) does not happen in nature. What is commonly called spontaneous generation is in fact produced by *semina* which are animated by the soul. *Semina* organise matter, which is completely inert. The fortuitous concourse of atoms cannot explain the generation of natural bodies. Even in the smallest atoms there are seminal principles which inform and give instructions to the passive matter.[91]

SENNERT, FREITAG, SPERLING

Sennert's *Hypomnemata* contain a short reply to Johannes Freitag, Groningen professor of medicine, a staunch anti-atomist and anti-

[87] Ibid., pp. 112; and 140-2.

[88] Cf. J. Roger, *Les sciences de la vie dans la pensée française au XVIIIe siècle* (Paris, 1993²), pp. 106-11, and W. Pagel, *New Light on William Harvey* (Basle and New York, 1976), pp. 84-92.

[89] Pagel (n. 88), p. 92.

[90] Sennert, *Hypomnemata* (n. 82), p. 139.

[91] Ibid., p. 420.

Paracelsian.[92] Freytag maintained that Sennert created a philosophical sect which took its inspiration from the teachings of Paracelsus and of Severinus. Following a common pattern, Freitag attacks the Paracelsians by resorting to what had become a reservoir of anti-Paracelsian arguments, namely Erastus's *Disputationes*.[93] The allegations are of impiety, blasphemy, immorality, of demonic magic, and of introducing obscure and meaningless terms in philosophy.[94] One of Freitag's targets is Sennert's theory that spirit and innate heat are not elemental, but originate from a celestial substance. Freitag thinks that this doctrine proves Sennert's adherence to Paracelsianim.[95] Moreover, Freitag believes that Sennert's theory of matter, which postulates the existence of atoms, is incompatible with Christian faith. The religious overtones of Freitag's polemic are also apparent from his criticism of Sennert's theory of the origin of forms and of souls. His view is that forms are educed from the potentiality of matter, and he overtly

[92] Johann Freitag (or Freytag) (1581-1641) studied in Helmstaedt, where he became professor of medicine and then physician to the bishop of Osnabrück. In 1631 he became Professor of Medicine at the University of Groningen. Freitag's arguments against Sennert are contained in the following works: *Aurora Medicorum Galeno-Chymicorum* (Frankfurt, 1630); *De Opii Natura* (Groningen, 1632); *Disputatio Medica de Morbis Substantiae, & Cognatis Quaestionibus Contra hujis tempestatis novatores & paradoxologos* (Groningen, 1632); *Disputatio Medico-Philosophica de Formarum Origine, quam adversus venerandae antiquitati repugnantem Neotericorum doctrinam Auditoribus suis exhibet Johannes Freytagius. Defendente eam Henrico Welman* (Groningen 1633); *Disputatio Medico-Philosophica prior De principiis rerum naturalium materialibus in genere, opposita Neotericorum quorandam & Pseudo-chymicorum... proponit Freytagius [praeses], Respondente Wilhelmo Henrico Cras* (Groningen, 1633). *Detectio et Refutatio novae sectae Sennerto-Paracelsicae, editio nova* (Amsterdam 1637, 1st edn: 1636). On Freitag see Partington, ii, p. 276.

[93] See Erastus, *Disputationum de medicina nova Philippi Paracelsi,* 3 parts, (Basle, 1571-3). On Thomas Erastus (1524-83) see Pagel, *Paracelsus* (n. 34), pp. 311-333; C.D. Gunnoe Jr., 'Thomas Erastus and his Circle of Anti-Paracelsians', in J. Telle (ed.), *Analecta Paracelsica. Studien zum Nachleben Theophrast von Hohenheims im deutschen Kulturgebiet der Frühen Neuzeit* (Stuttgart, 1994) pp. 127-48.

[94] Freitag, *Aurora Medicorum Galeno-Chymicorum: seu de recta purgandi methodo libri IV* (Frankfurt, 1630).

[95] Freitag, *Disputatio Medica de Calidi innati essentiam juxta veteris Medicinae & Philosophiae decreta explicans, opposita neotericorum & novatorum paradoxis, Praeses, Freitag, resp. Conrad Walter* (Groningen, 1632-3); and id., *Detectio et solida refutatio* (n. 92), pp. 14; 24-5.

rejects Sennert's position ("forma est sui multiplicativa") on the grounds that it undermines the unity of forms: "Forma non disponit sibi materiam, ut ea ingrediatur & habitet, sed materia disponitur ut formam recipere possit."[96] Contrary to the idea that the seed is animated, Freitag maintains that the soul exists only in organic bodies;[97] and that Sennert's impiety is also attested by his doctrine of the transmission of souls from the parents (*per traducem*), which would imply the transmigration of the souls, as well as the immortality of the soul of beasts.[98]

Sennert's very short reply to Freitag appears in the Preface to *Hypomnemata Physica*. Its purpose is to distance his own philosophy from Paracelsianism. The task of replying to Freitag in detail is in fact taken over by Sperling, Sennert's student at Wittenberg.[99] In his first reply to Freitag (1634), besides vindicating his mentor's theory of the origin of forms, Sperling unambiguously advocates the atomistic theory of matter: "Corpora in corruptionibus in multas atomorum myriades abire & in generationibus etiam ex atomorum innumero constari."[100] Atomism plays a central part in Sperling's *Institutiones Physicae*, a comprehensive textbook of philosophy, based largely on Sennert's teaching. Though Sperling does not rule out the notion of form, his natural philosophy has a stronger corpuscular bent than Sennert's. Sperling maintains that the four elements rank above the three chemical principles. But, he continues, mixed bodies are seldom decomposed into the four elements: upon analysis they yield the *tria prima*.[101] Unlike Sennert, Sperling (following Basso) explains the properties of salt, sulphur and mercury as the outcome of the textures of their atoms.[102]

[96] See *Disputatio medico-philosophica de formarum origine, Praeses Freitag, defend. Henricus Welman* (Groningen, 1633). See also *Disputatio Medica de Morbis Substantiae, & cognatis quaestionibus, Praeses Freitag, resp. Jacobus Martini* (Groningen, 1632).

[97] Freitag, *Detectio* (n. 92), Disp III, *Praeses* Freitag, *defend.* Henricus Magnus Heigel, p. 128.

[98] Ibid., pp. 126-36.

[99] Johann Sperling was born in Thuringia on 12 July 1603. He studied Theology and Medicine in Wittenberg, where he got his degree of M.D. in 1625. In 1634 he became Professor of Natural Philosophy in the same University.

[100] J. Sperling, *Tractatus Physico-Medicus De Origine Formarum* (Wittenberg, 1634), p.429.

[101] Sperling, *Institutiones Physicae* (Wittenberg, 1639), p. 869.

[102] Ibid., p. 857. For Basso, see below, chapter 2, pp. 39-42.

CONCLUSION

Sennert's corpuscular philosophy did not reject some crucial aspects of Aristotelian philosophy, such as the notion of substantial form – though he interpreted this notion in his own way. Sennert considered the mixed body (*mixtum*) as an aggregate of particles of elements, whose substantial forms are subordinated to the 'higher' form of the compound. Sennert's pluralistic and hierarchical view of forms made it possible for him to adopt the corpuscular theory within the framework of the Aristotelian theory of form. It must however be stressed that Sennert's forms are assisted by spirit, that is, the Neoplatonic astral body. Sennert's spirit was not identical with the Paracelsians', but certainly had much in common with it. It is apparent that Freitag's definition of Sennert as crypto-Paracelsian was not entirely groundless. Indeed, Sennert adopted the three chemical principles, though as subordinate to the four elements. He accepted chemistry as a practical discipline, yet he endeavoured to establish a line between chemistry and Paracelsian philosophy. Sennert's works (notably *De Chymicorum* and *Hypomnemata Physica*) contain the first articulate fusion of chemistry and corpuscular philosophy. Chemical analysis and synthesis are explained as the result of *synkrisis* and *diakrisis* of unchangeable corpuscles endowed with distinct chemical qualities. Atoms do not interact in a mechanical way, their movements being directed by a formative principle, which Sennert identifies with form and with spirit. Finally, Sennert's use of chemistry to prove the existence of atoms was one of the most influential features of his work. The 'qualitative' version of atomism was very common in the first half of the seventeenth century. Various natural philosophers adopted the corpuscular theory of matter, but they did not rule out the existence of elements and principles. Their corpuscles were not endowed with purely mechanical properties, but with qualities and powers. This was the case, for example, with Hill, de Boodt, van Goorle, Digby, Berigardus and Magnenus – whose works will be discussed in the ensuing chapters.

CHAPTER 2

SPIRIT, CHEMICAL PRINCIPLES AND ATOMS IN FRANCE IN THE FIRST HALF OF THE SEVENTEENTH CENTURY

INTRODUCTION

In France, as distinct from Germany, Paracelsianism and chemistry were up against a strong and enduring opposition, on both intellectual and political grounds.[1] The medical establishment launched a virulent attack on Paracelsian medicine – witness the antimony controversy. In the second half of the sixteenth century, Paracelsianism was becoming widespread among Paris physicians and, as Didier Kahn has shown, Roch le Baillif was by no

[1] For a general view of Paracelsianism, see A.G. Debus, *The Chemical Philosophy*, 2 vols, (New York, 1977); W. Pagel, *The Smiling Spleen. Paracelsianism in Storm and Stress* (Basle, 1984); H. Trevor Roper, 'The Paracelsian Movement', in *Renaissance Essays* (London, 1986), pp. 149-99; J. Telle (ed.), *Parega Paracelsica. Paracelsus in Vergangenheit und Gegenwart* (Stuttgart, 1991); id. (ed.), *Analecta Paracelsica. Studien zum Nachleben Theophrast von Hohenheims im deutschen Kulturgebiet der frühen Neuzeit* (Stuttgart, 1994); H. Schott und I. Zinguer (eds.), *Paracelsus und seine internationale Rezeption in der frühen Neuzeit. Beiträge zur Geschichte des Paracelsismus* (Leiden, 1998); O.P. Grell (ed.), *Paracelsus: the Man and his Reputation, his Ideas and their Transformation* (Leiden, 1998). For the French scene, see Metzger, *Doctrines Chimiques*; H. Guerlac, 'Guy de La Brosse and the French Paracelsians', in A.G. Debus (ed.), *Science, Medicine and Society in the Renaissance. Essays to Honor Walter Pagel*, 2 vols. (New York, 1972), i, pp. 177-85; H. Trevor-Roper, 'The Sieur de la Rivière', in *Renaissance Essays*, pp. 200-22; A.G. Debus, *The French Paracelsians. The Chemical Challenge to Medical and Scientific Tradition in Early Modern France* (Cambridge, 1991); and Didier Kahn, 'Paracelsisme et alchimie en France à la fin de la Renaissance (1567-1625)' (unpublished dissertation, Paris IV, 1998).

means alone.[2] The supposed association of alchemists and Paracelsians with the Rosicrucians contributed towards dramatising the controversies over Paracelsianism.[3]

Chemistry was taught privately in Paris by Jean Beguin, whose *Tyrocinium Chymicum* (1610) became one of the most popular chemical textbooks in the seventeenth century. The institutionalisation of chemistry was due to Théophraste Renaudot and Guy de la Brosse. The latter was responsible for the foundation of the Jardin du Roy, where chemistry was regularly taught.[4] He also wrote a treatise on chemistry, subscribed to the *tria prima* theory and unambiguously rejected the Aristotelian elements.[5]

Though historians of chemistry and alchemy have paid little attention to the relationship between French chemistry and the corpuscular philosophy, it is apparent that these two traditions were closely linked in the first half of the century. Indeed, the standard view (held by historians of science and of philosophy) is that the corpuscular theory of matter, which led to Gassendi's and Descartes's mechanism, developed in opposition to chemical philosophy, and in particular to the Paracelsian doctrine of principles.[6] In the present chapter I argue: 1. that corpuscular notions are to be found in a number of alchemical and chemical treatises; 2. that the works of Basso, Gassendi, and Mersenne show at different degrees the influence of alchemical and chemical theories. Descartes's mechanical philosophy,

[2] D. Kahn, 'La Faculté de Médecine de Paris en échec face au Paracelsisme: enjeux et dénouement réels du procès de Roch le Baillif', in Schott and Zinguer (n. 1), pp. 146-221.

[3] On the 'antimony war', see P. Pilpoul, *La querelle de l'antimoine* (Paris, 1928) and A.G. Debus, *French Paracelsians* (n. 1), pp. 21-30.

[4] On Renaudot see H.M. Solomon, *Public Welfare, Science and Propaganda in Seventeenth-Century France: The Innovations of Théophraste Renaudot* (Princeton, 1972). On Guy de la Brosse see DSB, Guerlac (n. 1); and id., 'Guy de la Brosse: Botanist, Chemist, and Libertine', in H. Guerlac, *Essays and Papers in the history of Modern Science* (Baltimore and London, 1977), pp. 440-50; R.C. Howard, 'Guy de La Brosse and the Jardin des Plantes', in H. Woolf (ed.), *The Analytic Spirit. Essays in the History of Science in Honor of Henry Guerlac* (Ithaca and London, 1981), pp. 195-224; id., *La bibliothèque et le laboratoire de Guy de la Brosse au Jardin des Plantes à Paris* (Geneva, 1983).

[5] Following Petrus Severinus, de la Brosse conceived water and earth as matrices and receptacles of the *semina rerum.* See G. de La Brosse, *De la Nature, Vertu et Utilité des Plantes* (Paris, 1628), pp. 289-440.

[6] Metzger, *Doctrines Chimiques*, p. 233; R. Lenoble, *Mersenne ou la naissance du mécanisme*, (Paris, 1943) pp. 134-153 and Debus (n. 1), pp. 154-155.

based as it is on the identity of matter and extension, marks a rupture with the previous corpuscular theories. In Descartes's works chemical phenomena are interpreted in mechanical terms, as if produced by the motions of particles of matter having different shapes and sizes. Since historians of science have considered early seventeenth-century atomism as preparatory to the mechanical theory of matter, relevant aspects of French atomism – which were not mechanical – have either been ignored or been explained as signs of the persistence of out-dated theories in the atomistic philosophy.

UNIVERSAL SPIRIT, CHEMICAL PRINCIPLES AND ATOMS

The doctrine of spirit played a substantial part in seventeenth-century natural philosophy and medicine. The Neoplatonic spirit of the world was widely adopted by chemical philosophers as a principle of motion and life. In a hierarchically organised universe it was deemed to be a substance originating in the stars and therefore superior to the four elements. This notion became central to chemistry and medicine thanks to Ficino's *De vita* and was widely adopted in Severinus's influential *Idea Medicinae Philosophicae* (1571). As William Harvey noticed in his *Exercitationes duae*, it was often employed as *factotum,* in both natural philosophy and medicine.[7]

Both in Duchesne's *Ad Veritatem hermeticae medicinae* (1604) and in d'Espagnet's Neoplatonic cosmology universal spirit and *semina rerum* play a central part. The forms of bodies have celestial origin and are contained in the seeds, which receive the formative power from the universal spirit – a substance diffused throughout the universe, giving life and activity to all bodies.[8] As Betty Dobbs pointed out, d'Espagnet postulated the existence of magnets in the air – that is, intermediate substances between spirits and bodies, able to attract a portion of the universal spirit. As we shall see,

[7] W. Harvey, *Exercitationes duae Anatomicae de Circulatione Sanguinis ad Joannem Riolanum filium* (Rotterdam, 1648), pp. 66-7.

[8] J. d'Espagnet, *Enchiridion Physicae Restitutae* (Paris, 1608). References are taken from the 1642 Paris edition, pp. 131, 137, 147-8. On Jean d'Espagnet (1564-1637), see Ferguson, *Bibliotheca Chemica*; F. Secret, "Littérature et alchimie, X: Mlle de Gournay alchimiste", *Bibliothèque d'Humanisme et Renaissance*, 35 (1973), 526-531; T. Willard, 'The Many Worlds of Jean d'Espagnet', in A.G. Debus and M.T. Walton (eds), *Reading the Book of Nature. The Other Side of the Scientific Revolution* (Kirksville, Missouri, 1998), pp. 201-14.

Boyle paid special attention to these 'magnets'.[9] Following Severinus, d'Espagnet maintained that natural phenomena are produced by incorporeal agents, bodies being like their external envelopes (*velut cortices*). As he put it, natural bodies do not act "per solas qualitates, ut vulgo placet, sed per Spiritus secretos".[10] Though d'Espagnet's view of nature is Neoplatonic, embryonic corpuscular views are employed to account for the *vexata quaestio* of the origin of mixed bodies:[11]

> Democriti sententiam, omnia corpora ex Atomis fieri, a Natura alienam non esse licet affirmare; eum enim ratio & experientia à calumnia vindicant: elementorum quippe mixtionem obscuro sermone velare, nec reticere omnino voluit ingeniosus Philosophus, quae ut Naturae intentioni congruat, per minima & actu indivisibilia corpuscula fieri necesse est, secus in corpus continuum & naurale non coalescerent elementa. Docet nos experientia in artificiali mixtorum resolutione & compositione, quae per distillationes exercentur, perfectam duorum aut plurium corporum mixtionem non fieri, nisi in subtili vapore: At longe subtiliores mixtiones suas & quodammodo spirituales facit Natura, quas Democritus intellexisse sanius credendum est: corporum enim crassities mixtioni obstat, propterea quanto magis res attenuantur, tanto mixtioni aptiores fiunt.[12]

According to d'Espagnet, the four elements are not the ultimate constituent of bodies. They are made of insensible corpuscles. However, the role of the elements in nature is by no means marginal. The chemical principles are in fact formed from the combination of the four elements (mercury from earth and water, sulphur from earth and air, salt from air and water). Fire is given a special status, being the material formative principle in all mixed bodies.[13]

Like d'Espagnet and Sendivogius, Clovis Hesteau de Nuysement's *Traittez de l'Harmonie et Constitution generalle du vray sel* resorted

[9] Cf. B.J.T. Dobbs, *The Foundations of Newton's Alchemy, or The Hunting of the Greene Lyon* (Cambridge, 1984), pp. 37-9, 153. A source of d'Espagnet's view of universal spirit and magnets might be M. Sendivogius's *Novum Lumen Chymicum* (Prague, 1604). On Sendivogius, see DSB. As we shall see in chapter 4, the quest for the celestial magnet crops up in Boyle's Correspondence of 1659.

[10] D'Espagnet (n. 8), p. 169.

[11] Cf. Lasswitz, *Geschichte*, i, pp. 333-9. d'Espagnet's atomism is ignored by Dobbs and by Debus. The latter labelled d'Espagnet's *Enchiridion* a "mystical writ", *French Paracelsians*, (n. 1), p. 177.

[12] D'Espagnet (n. 8), pp. 122-4.

[13] Ibid., pp. 44-5; 119-20.

heavily to the universal spirit. Yet, unlike d'Espagnet's *Enchiridion*, the *Traittez* does not contain corpuscular views. The world, for Nuysement, is a living body, having soul and spirit. The latter is the vital principle contained in the seeds of things and is to be found (at different degrees) in all natural bodies – the four elements being almost deprived of any role and subsidiary to the spirit.[14] Similarly, Henry de Rochas makes the universal spirit the centre of his natural philosophy. Matter being passive, the principle of activity is deemed to be the spirit of the world. De Rochas's views of the constituents of bodies are close to those of the Paracelsians, as he unambiguously rejects the Aristotelian elements and adopts the spagyrical principles.[15] Yet water has a special status as it corporifies the universal spirit and impregnates it.[16] De Rochas's theory of chemical principles is somewhat articulated, as he distinguishes between three kinds of salt, i.e., fixed, volatile and nitrous; and shows the difference between salts extracted from plants and those obtained from animals.[17]

ATOMS AND PRINCIPLES: BASSO

Basso's *Philosophia Naturalis* (1621) stands out as one of the earliest and most articulate expositions of the corpuscular theory of matter. Since the place of Basso's *Philosophia* in the history of atomism has been dealt

[14] Nuysement, *Traittez de l'Harmonie et Constitution generalle du vray sel, secret des Philosophes, & de l'Esprit universel du Mond* (Paris 1621), pp. 2; 11; 17; 23. On Nuysement see W. Kirsop, 'Clovis Hesteau, sieur de Nuysement, et la littérature alchimique en France à la fin du XVIe et au début du XVIIe siècle' (unpublished dissertation, Université de Paris, 1960). As S. Matton ('La figure de Démogorgon dans la littérature alchimique', in D. Kahn and S. Matton, *Alchimie: art, histoire et mythes* (Paris and Milan, 1995), pp. 308-17), and D. Kahn (n. 2) have demonstrated, Nuysement's *Traittez* is entirely based on Jean Brouaut's *Trois livres des elemens chymiques et spagyriques*.

[15] H. de Rochas, *La Physique Reformée, contenant la refutation des erreurs populaires, et le triomphe des veritez philosophiques*... (Paris, 1648, first edn: 1638), pp. 60, 133; id., *La Physique demonstrative*, (Paris, 1644), pp. 148-9; id., *Histoire des eaux minerallες* (Paris, 1648), 2 parts with separate pagination, part i, pp. 31, 234, 239-51. On Rochas (1619-1648) see Ferguson, *Bibliotheca Chemica* and S. Matton, "Henry de Rochas plagiaire des 'Trois livres des elemens chymiques et spagyriques' de Jean Brouaut", *Chrysopoeia* 5 (1992-1996), 703-719.

[16] Rochas, *La Physique* (n. 15), p. 60.

[17] Rochas, *Histoire des eaux minerallles* (n. 15), vol. i, pp. 174-6 and vol. ii (separate pagination), p. 6.

with in a number of studies, here I focus on the relationship of atomism to the Paracelsian chemical philosophy in Basso's work.[18] Unlike Sennert, Basso unambiguously rejected the Aristotelian notion of form as obscure and superfluous. For Basso, the different arrangements and the local motions of unchangeable corpuscles explain all natural phenomena.[19] Motion is not a property of atoms, it originates from spirit – which Basso identifies with the world soul – and ultimately from God.[20]

There is however evidence that Basso's atomism did not encompass the mechanisation of nature. When Basso speaks of motion he also means *appetitus*, attraction and repulsion, sympathies and antipathies.[21] Basso's corpuscles are not endowed with mechanical properties (shape and size), but have specific *naturae*, which remain unchanged in the *mixtum*.[22] When Basso refers to corpuscles having different natures he means particles of the four elements or the five chemical principles. He maintains that chemical analysis shows that the chemical principles (salt, sulphur and mercury, plus earth and phlegm) are extracted from all natural bodies.[23] Basso also

[18] S. Basso, *Philosophiae Naturalis adversus Aristotelem libri XII* (Amsterdam, 1649, first edn Geneva 1621). On Sebastien Basso's life, see DBI and C. Lüthy, 'Thoughts and Circumstances of Sébastien Basson. Analysis, Micro-history, Questions', *Early Science and Medicine* 2/1 (1997), 1-73. On Basso's philosophy, see Lasswitz, *Geschichte*, i, pp. pp.467-81; T. Gregory, 'Studi sull'Atomismo del Seicento, I', *Giornale Critico della Filosofia Italiana* 44 (1964), 38-65; G. Zanier, 'Il macrocosmo corpuscolaristico di Sebastiano Basson', in *Ricerche sull'Atomismo del Seicento* (Firenze, 1977), pp. 77-118, L.O. Nielsen, 'A Seventeenth-Century Physician on God and Atoms: Sebastian Basso', N. Kretzmann (ed.), *Meaning and Inference in Medieval Philosophy* (Dordrecht, 1988), pp. 297-369.

[19] "Caeterum quomodo ex illis diversissimis particulis primis partes in infinitum discrepantes conflari possint; atque per aliquarum particularum, vel detractionem, vel additionem, vel situs partium variationem, aliae in aliarum naturam facile transeant, non intellectu est difficile.", Basso (n. 18), pp. 72; 118 and 387.

[20] Ibid., p. 302. Nielsen stresses that the motion of ether, and consequently of atoms, completely depend on God's will. See Nielsen (n. 18), pp. 318-23 and 343-4.

[21] Basso (n. 18), p. 391: "Diximus spiritum illum universalem elementis coniunctum duplicem illis appetitum impertiri, similis scilicet coniunctionem, et spacium, locumque suae naturae debitum. Primus ille appetitus duos motus excitat, unum primarium quo scilicet simile attrahit simile, vel forte potius ad simile se recepit; neque enim vi fit talis attractio, sed amore. Alterum secundarium quo scilicet dum simile appetit similis connexionem"

[22] Ibid., p. 112.

[23] "Vix ulla res est ex qua non eliciant tres naturas valde inter se differentes; quarum quae subtilior est et volatilior, spiritum vocant, seu etiam mercurium, Quae

mentions corpuscles which are prior to the elements and principles, but he is rather vague about their properties.[24] His reluctance to give a clear definition of atoms is attested by the fact that he does not even choose between the Platonic two-dimensional elementary parts and the Democritean solid atoms. With the exception of fire particles, Basso never refers to the shape and size of corpuscles. He only asserts that the simplest particles are indivisible and unchangeable. However, for Basso, the analysis of natural bodies does not always yield their ultimate constituents, namely the five principles. What is often obtained is a complex substance.[25] This view of Basso's is based on his classification of substances – and accordingly of corpuscles – which was adopted by Sennert and thoroughly developed by Robert Boyle. The classification of corpuscles is stated in the following terms:

> Non tantum prima elementa in misto seu mavis composito manere, sed diversissimas quibus mistum constat, particulas, ex iis primis rerum principiis diversimode constructas; quas secundas, docendi gratia vocare liceat. Ex his secundis, varie coëntibus, tertiae fiunt non minus quam secundae inter se differentes. Eundem in modum, & ex tertiis quartae, & ex quartis quintae fieri intelliguntur. Ad cujus rei confessionem nos vel invitos ducit experientia in rei cujusvis dissolutione. Quippe, res in ea resolvitur, ex quibus componitur. Atqui compositum naturale non primo resolvitur in prima illa elementa, sed in partes quasdam inter se natura discrepantes; quarum singulae species rursus in alios multiplices dividuntur; & hae in alias minutiores conciduntur. Saepiusque haec partium diversarum in minutiores semper diversas subdivisio repetitur.[26]

vero crassior et pinguior, oleum et sulphur appellant, Quae vero omnium maxime fixa ex intimis veluti partis cuiusque penetralibus ultima educitur, sal illis nuncupatur. Praeter has tres naturas valde utiles, superest quaedam materia terrestris et inutilis quam foeces vocant et caput mortuum, est insuper aquaeus quidam et insipidus liquor quae phlegma dicunt." Ibid., pp. 31-2.

[24] "Materia rerum ex minutissimis particulis diversae naturae comparata est; quae quidem naturae sive sint quatuor elementa: ignis, aër, aqua, terra; sive quid aliud prius, ex quo haec elementa componantur, speciei diversissimae sunt. Caeterum, naturas illas, quae ad ignem conficiendum sunt magis idoneae, nos ignem vocamus, & ita de caeteris." Ibid., p. 112.

[25] Ibid., p. 70.

[26] Ibid., p. 70. Cf. H.H. Kubbinga, 'Les premières théories 'moleculaire': Isaac Beeckman (1620) et Sébastien Basson (1621). Le concept d'*individu substantiel* et d'*espèce substantielle*', *Revue d'Histoire des sciences* 37 (1984), 215-33.

While the primary corpuscles or atoms, which are so small that they cannot affect our senses, are indestructible and always remain unchanged, those of second, third and greater degrees of composition may decompose and change their natures in the course of the chemical reactions.[27] A closer investigation of Basso's natural philosophy reveals the presence of a number of Paracelsian elements. Besides adhering to the doctrine of chemical principles, Basso adopts a cosmology which is indebted to Paracelsus. Like Cardano, Basso maintains that celestial influences give life and perfection to terrestrial bodies. In addition, Basso believes that a continuous process of concoction, digestion and purification of parts of fire occurs in celestial bodies.[28] This is not asserted as a mere analogy. Basso, like Paracelsus and some of his followers, believes that physiological processes take place in the celestial bodies.[29]

ETIENNE DE CLAVE

As Hélène Metzger pointed out, de Clave occupies an intermediate position in seventeenth-century French chemistry. Unlike the majority of Paracelsians, de Clave does not build up a cosmology based on chemistry. Yet, neither does he confine chemistry to a purely technical discipline – as Jean Beguin did.[30] His chemical works deal extensively with the doctrine of principles, besides containing many experiments and observations both in chemistry and in mineralogy. De Clave is well known for the fourteen theses of 1624, which produced a censure from the Sorbonne, an *arrêt* of Parliament (4 September 1624) and two refutations, one written by Jean Baptiste Morin (which also contains the text of the fourteen theses), and one by Mersenne.[31] De Clave developed Duchesne's spagyrical doctrine of the

[27] "Hae primae particulae adeo minutae sunt, ut nisi plurimae in unam molem coeëant, sensum non afficiant. [...] Hinc licet primae sint immutabiles quantum ad essentiam, secundae & tertiae & reliquae facile possunt aliae in aliarum naturam transire, paucioribus, aut pluribus mutatis, prout magis vel minus aliae ab aliis recedunt." Basso (n. 18), p. 113.

[28] Ibid, p. 304. Cf. Zanier (n. 18), pp. 104-110.

[29] According to Basso, air contains "varia rerum semina", Basso (n. 18), p. 21.

[30] Metzger, *Doctrines chimiques*, pp. 53-4. On Beguin, see T.S. Patterson, 'Jean Beguin and his Tyrocinium Chymicum', *Annals of Science* 2 (1937), 243-98.

[31] J-B. Morin, *Refutation des theses erronees d'Antoine Villon..., & Etienne de Claves*...(Paris, 1624). Mersenne, *Verité des Sciences*, (Paris, 1625), pp. 79-83. See D. Kahn, 'Entre atomisme, alchimie et théologie: la réception des thèses d'Antoine

chemical principles and adopted Ficino's and Severinus's notion of spirits and *semina* as the main agents in nature. When he dealt with the problem of the *mixtio,* de Clave made use of the notion of *minima naturalia*, though he did not develop it into an articulate corpuscular theory of matter.

The impact of the 1624 theses was certainly due to their unambiguous rejection of fundamental Aristotelian notions, like prime matter, privation, substantial forms and the four elements. Though very short, the theses explicitly aimed to replace the Aristotelian theory of elements with a combination of the five-principle doctrine and the atomic theory of matter:

> Thesis V: Car le mixte est composé de cinq corps simples ou elemens, existans en luy actuelement & formellement, sçavoir de Terre, d'eau, de sel, de soufre ou huile & de Mercure ou esprit acide: qu'on doit estimer les vrais & seuls principes naturels; comme ceux qui ne sont faicts ny d'eux-mesmes, ny d'autres choses, mais desquels sont faits tous les composez naturels. Thesis VI: Ces principes sont ingenerables & incorruptibles, & de mesme espece infime dans tous les mixtes. Thesis XIV: Par toutes ces choses, il est tres manifeste que ces deux dits des Anciens, Toutes choses sont en toutes choses, & toutes choses sont composees d'atomes ou indivisibles, on esté ignoramment ou plutost malicieusement bafouez par Aristote. Et parce que l'un & l'autre est conforme à la raison, à la vraye philosophie, & à l'anatomie des corps nous le deffendons obstinément, & soustenons fermement.[32]

As they are described in thesis XIV, de Clave's corpuscles are a compromise of Democritus's atoms and Anaxagoras's homeomeries, i.e. a qualitative, non-mechanical atomism.[33] De Clave's qualitative version of the corpuscular philosophy is evidently linked to the five-principle theory. Atoms are in fact interpreted as the smallest units of the five chemical principles. The rejection of the transmutations of elements into each other (thesis XIII) is based on the assumption that corpuscles of water, earth, salt and sulphur are the basic units of all mixed bodies and cannot be split or transformed into different substances.

de Villon et Étienne de Clave contre Aristote, Paracelse et les "cabalistes" (24-25 août 1624)', *Annals of Science*, forthcoming.

[32] Morin (n. 31), pp. 13-7.

[33] Ibid., p.17.

De Clave's first printed book is *Paradoxes ou traittez philosophiques des pierres et pierreries*, published in 1635, a work devoted to mineralogy.[34] Here de Clave placed special emphasis on the role of spirit as the agent of generation. For de Clave, stones are not produced from the four elements or from the chemical principles. He explains their origin by having recourse to seminal principles, which contain spirit, i.e., a formative principle.[35] The action of spirit is described in physical terms. A very subtle substance, spirit unlocks the most compact elements and gives access to the smallest particles.[36] It produces the combinations of the *minima* of the five principles and also the *mixtum*:

> alors l'esprit agité par cette chaleur exterieure, pestrit le tout à cause qu'il se communique & s'insinuë le plus facilment de tous avec les autres (comme nous avons declaré en son traitté particulier) & fait qu'estants attenuez & subtiliez, ils s'incorprent ensemble per minima; c'est à dire estroittement, & par tres menues & subtiles parcelles, tant que faire se peut.[37]

De Clave's definition of spirit contains some degree of ambiguity, as he describes this substance both as distinct and more active than the five principles and also as an acid spirit, namely, one of the five principles.

In the *Nouvelle lumiere philosophique des vrais principes et elemens de nature*, which appeared in 1641, de Clave articulated his view of the five principles by criticising the traditional spagyrical theories. His main point was that if a given substance is simple and homogeneous, it does not entail its being an element (or principle). In order to achieve that status, a given

[34] E. de Clave, *Paradoxes ou traittez philosophiques des pierres et pierreries* (Paris, 1635). It would seem that the author had already written a number of tracts on mineralogy, chemistry and medicine before 1635. See 'Preface' sig. Eiij$^{r\text{-}v}$.

[35] "Nous disons donc que ce n'est pas le lieu qui donne la faculté generative à la semence, ouy bien l'esprit qui est contenuu en icelle, excité au prealable par l'agent externe..." E. de Clave, *Paradoxes* (n. 34) p. 346. See also ibid. pp. 366-7.

[36] "Car cet esprit seminaire, comme nous dirons plus amplement cy-aprés, est la vraye semence qui ouvre les elemens plus compactes, pour donner entrée en iceux aux autres moin grossiers, & qui neantmoins n'y pourroient avoir aucun accés sans cét esprit, qui ouvre et mesle toutes les autres substances diverses & heterogenes, pour les rendre comme homogenes, et les unir sous une mesme forme." Ibid., p. 368.

[37] Ibid., p. 225.

chemical substance has to be the component of all mixed bodies.[38] In de Clave's classification of chemical principles, the status of spirit is again rather ambiguous. Unlike d'Espagnet and Basso, he states that the chemical spirit is not to be confused with the astral body. He distinguishes the chemists' spirits from the spirits of the theologians and physicians, and includes it among the principles. However, de Clave defines spirit in two ways which might appear to be incompatible. Spirit is defined both as one of the chemical principles and also as energy informing all the principles. As a corporeal substance, it is the most subtle and penetrating of the principles.[39] As energy, spirit is "comme une vertu qui agit en tout diversement, suivant la nature particuliere de chaque principe, sans se mesler avec iceluy."[40] It is a mostly subtle substance, which has the power of producing fermentation, one of nature's most relevant operations. Fermentation, for de Clave, is "une attenuation des parties plus crasses & grossieres", which brings about a new composition.[41]

De Clave's description of the remaining four principles follows a traditional pattern. The only original view to be found in his work is that volatile salts are not considered as simple chemical principles, but as compound bodies, being made of salt and spirit.[42] He adopts a quantitative approach to the chemical principles and classifies the five elements according to their different specific weights: the lightest is oil, then earth, water, salt and finally the heaviest, spirit.[43] However, de Clave does not accept the mechanical explanation of qualities, like the one to be found in Galileo Galilei's *Il Saggiatore* (1623). De Clave rejects the view that heat

[38] "Nous disons donc qu'il y a cinq corps simples, que nous appellons elemens, non pas à cause qu'ils sont simples: autrement le Ciel & l'air seroient elemens, ains seulment par ce qu'ils composent tous les mixtes." E. de Clave, *Nouvelle lumiere philosophique des vrais principes et elemens de nature* (Paris, 1641), p. 159.

[39] "Il faut donc sçavoir que les Chymistes veulent que ce [l'esprit] une substance corporelle, la plus subtile & penetrante du mixte, laquelle estant liberée des liens d'iceluy, ouvre, dissout, penetre ou permée les corps mixtes, voire le plus compactes, pour ayder à la separation des diverses, voire plus pures parties du mixte." Ibid., p. 65.

[40] Ibid., p. 68.

[41] Ibid., pp. 46-7; 55. On fermentation in seventeenth-century chemistry and medicine, see W. Pagel, *Jan Baptista van Helmont. Reformer of Science and Medicine* (Cambridge, 1982), pp. 79-87.

[42] E. de Clave, *Nouvelle Lumiere* (n. 38), pp. 40-1.

[43] Ibid., p. 101. It is somewhat surprising to see that spirit is both the most active and the heaviest principle. It is apparent that de Clave identifies spirit with mercury.

does not exist *per se*, but is the result of the motion of particles and of their encounter with our organs of sense: heat, he says, has objective existence, as it belongs to the three active elements, i.e., spirit, salt and oil:

> Voila quant à la chaleur & à la froideur; mais si nous les considerons suivant l'attouchement qui est appellé ou defini par quelques-uns, un mouvement des parties tres menuës de la matiere reverberée en soy-mesme, penetrant & dechirant le tact, comme par mille & mille pointes, nous trouverons que ceux-là se sont lourdement trompez, qui ont voulu exclure les elemens, d'autant que l'huile, l'esprit & le sel produisent les mesmes effects à nostre attouchement, comme nous venons de declarer aux articles precedens: par ainsi nous pouvons dire que la chaleur ne laisse pour cela d'estre une qualité fixe en l'huile, en l'esprit, & au sel, & un mouvement en iceux à nostre attouchement. Et de plus, ces gens là errent de dire qu'il ne se trouve parmi nous aucun corps qui soit perpetuellement chaud, puis que l'huile, l'esprit, & le sel ne peuvent iamais estre refroidis tandis qu'ils demeurent en leur simplicité, & sans meslange.[44]

However, de Clave does not entirely dismiss the kinetic theory of heat: he recognises that the particles of spirit, salt and oil are in motion and operate by means of their continuous movement. The difference with the mechanical theories is that for de Clave heat is a primary, not a secondary quality.[45] Like the *Paradoxes* of 1635, the *Nouvelle Lumiere* contains some rudiments of the corpuscular theory of matter, that is very close to the *minima naturalia* tradition. This is expressed in the context of de Clave's

[44] Ibid., p. 222.

[45] "De ce que dessus nous pouvons inferer que la chaleur & la froideur sont bient un mouvement, & de plus sont qualitez tactiles, la chaleur, dautant qu'elle congrege & assemble les choses homogenes & de mesme nature, & separe les heterogenes ou dissimilaires, & en outre est une qualité qui affecte nostre attouchement, en sorte qu'elle separe tant qu'elle peut, & suivant sa chaleur plus ou moins grande, les choses heterogenes, premierement les plus volatiles, & en suite celles qui le sont moins, & neantmoins elle ne laisse de causer un mouvement de parties reverberé en soy, parce qu'elle se met avec reverberation & prompte alteration: mais la difference qu'il y a de la chaleur de nos elemens à celle qui se fait par la reflexion & repercussion de la lumiere & de la collision reïterée des corps compactes, consiste en ce que toutes ces choses eschauffent par le seul mouvement; & nos trois elemens chauds, huile, esprit, & sel, produisent le mesme effect, non seulement par le mouvement, mais encores par leur qualité de chaleur, qui leur est tellement inherente & fixe, qu'ils ne peuvent recevoir aucune qualité contraire…" Ibid., pp. 223-4.

rejection of the Collegium Conimbricense's view of the composition of mixed bodies.[46] The union *per minima*, claims de Clave, is a physical process which produces a perfect mixture of the elements.[47]

De Clave's *Cours de Chimie*, which was published (and possibly prepared for the press on the basis of students' notes) by Olivier de Varennes in 1646, adds nothing to the views expressed in his previous works. The importance of de Clave's *Cours* – and of his works in general – lies in the fact that chemistry is not confined to giving practical instruction, but deals with a number of theoretical issues such as the principles of mixed bodies and the *minima naturalia*. De Clave contributed to accord chemistry a respectable position in French natural philosophy and influenced the work of Gassendi.

MARIN MERSENNE AND HIS CORRESPONDENTS

Mersenne's rejection of Paracelsian doctrines did not entail the dismissal of chemistry as a whole, which in fact he saw as a valuable contribution to the study of natural philosophy. As historians have stressed, Mersenne's criticisms were directed mainly against the alchemists' and the Paracelsians' fusion of alchemy with religion, notably their claim to produce a chemical interpretation of Creation.[48] This is apparent in Mersenne's head-on attack against Fludd, and in his rejection of several aspects of Paracelsian magic, as for example the view of imagination as an active power and the 'sympathetic' cure of wounds.[49] It must be stressed that the atomists,

[46] On the Conimbricenses' commentaries, see D. des Chene, *Physiologia. Natural Philosophy in Late Aristotelian and Cartesian Thought* (Ithaca and London, 1996), *passim*.

[47] E. de Clave, *Nouvelle Lumiere* (n. 38), pp. 275-6.

[48] On Mersenne see Lenoble, *Mersenne* (n. 6); and P. Dear, *Mersenne and the Learning of the Schools*, (Ithaca-New York, 1988). On Mersenne and alchemy see Debus, *French Paracelsians* (n. 1), pp. 72-3; and A. Beaulieu, 'L'attitude nuancée de Mersenne envers la chymie', in J.-C. Margolin and S. Matton (eds.), *Alchimie et philosophie à la Renaissance* (Paris, 1993), pp. 395-403.

[49] M. Mersenne, *Quaestiones Celeberrimae in Genesim* (Paris, 1623), cols. 539; 565-6. Besides attacking Paracelsus and Fludd, Mersenne selected Khunrath as a champion of the impious alchemy and magic. In the *Questions Théologiques, Physiques, Morales et Mathematiques* (Paris, 1634), pp. 133-4, Mersenne published, with approbation, the Sorbonne's Censure of Heinrich Khunrath's *Anphitheatrum Sapientiae* (1609), which reads: "La Sacrée Faculté de la Theologie de Paris, à tous les Catholiques Puisque l'Apostre nous enjoint d'éprouver toutes choses, & de

particularly Basso, van Goorle, Hill and the anti-Aristotelian Nathanael Carpenter, did not escape Mersenne's criticism. Like the alchemists, the atomists were accused of impiety.[50]

Even if in the *Quaestiones Celeberrimae in Genesim* Mersenne's opposition to the *neoteroi* (including Paracelsus, Croll and Campanella) was particularly virulent, he did not rule out the importance of the quest for the philosophers' stone, and advocated the foundation of an academy with the scope to co-ordinate the work of the alchemists – a suggestion which he repeated in *La Vérité des Sciences*, where he also stressed the importance of reforming the chemical terminology.[51] The duty of chemists is formulated in *Questions Inouyes* (1634), where Mersenne maintained that chemists should not pretend to frame a cosmology or to establish the foundation of medicine. Their task should be confined to the writing of faithful reports of their experiments.[52]

retenir ce qui est bon, ayant apperceu que depuis quelques mois les Catholiques ont un certain livre tres-pernicieux entre les mains, dans lequel il y a premierement quelques figures, & puis plusieurs explications de divers passages de la saincte Escriture disposees par sept degrez, & finalement quelques corollaires, & dont le titre est L'Amphiteatre Christianocabalistique Divinomagique... la sudite Faculté de Theologie yant leu exactement, & examiné le livre entier par quelques docteurs qu'elle a specialement deputez pour ce sujet, a jugé que les explications estant prises à la lettre, & tous les corollaires pris comme ils sont, avec le livre mesme, doivent estre condamnés, particulierement parce qu'estant remply d'ipietez, d'erreurs, & d'Heresies, & d'une perpetuelle profanation sacrilege des passages de la saincte Escriture, il abuse de plus saints mysteres de la Religion Catholique, & conduit les lecteurs aux arts deffendus & abominables. c'est pourquoy elle a jugé qu'un livre si contageieux ne peut pas estre leu, ny exposé en public sans perte de la Foy, de la Religion, & de la pieté.", dated March 1, 1625.

50 M. Mersenne, *L'Impieté des Déistes* (Paris, 1624), pp. 238-9; id., *Quaestiones Celeberrimae* (n. 49), col. 1838. In *La Verité des Sciences* (Paris, 1625), pp. 78-83, Mersenne's target was the theses of 1624. On van Goorle, see below, pp. 184-5; on Nicholas Hill, see below, pp. 75-7. See Nathanael Carpenter, *Philosophia Libera* (London, 1621). On Carpenter see DNB and C.B. Schmitt, 'Nathanael Carpenter', in F. Ueberweg, *Grundriss der Geschichte der Philosophie*, revised edn (gen. ed. R.W. Meyer), *Die Philosophie des 17. Jahrhunderts, 3: England*, ed. J-P. Schobinger (Basle, 1988), pp. 355-6 and 488.

51 Mersenne, *Quaestiones Celeberrimae* (n. 49), 'Praefatio' and col. 1483; id., *La Verité des Sciences* (n. 50), pp. 105-6.

52 Mersenne, *Questions Inouyes, ou Récréation des Sçavans* (Paris, 1634), p. 126.

Mersenne's correspondence with chemists (and in particular with van Helmont) testifies to his concern for the development of a 'purified' chemistry, i.e., chemistry and alchemy freed from the Paracelsians' religious and philosophical overtones. The letters to Mersenne (unfortunately the other side of the correspondence has been lost) contain evidence of the strong links between chemistry and atomism in seventeenth-century France.

MERSENNE ON THE CONSTITUTION OF MIXED BODIES

According to Mersenne's probabilistic theory of knowledge, physical sciences (including the theory of matter) cannot achieve absolute certitude.[53] The only certainty about the constitution of bodies, he believes, is in the Scriptures, where we read that water and earth are the two elementary substances. Therefore, fire may not be included among the elements.[54]

He adopted the corpuscular theory of matter and criticised the chemical principles along the following lines. First, he claimed, the *tria prima* are not the simplest and ultimate ingredients of bodies, as they are generated by heat and in addition can be decomposed by fire. Second, the chemists fail to explain why each principle is endowed with a given set of properties and no other. According to Mersenne, the origin of the properties commonly associated with the *tria prima* is not explained by the chemists. As we shall see, some of Mersenne's objections to the chemical doctrine of principles will be endorsed by Robert Boyle.[55]

Chemical operations are discussed in *Questions Théologiques*... (1634), where Mersenne takes into account the role of salt in the generation of metals and thoroughly discusses Rey's experiment of calcination of tin – which will be dealt with in the next paragraph. Mersenne accepts the view that salts play some role in the generation of minerals by producing cohesion, but he stresses that our knowledge of the different properties of salts is still inadequate and little can be assessed about their operations in nature.[56] On the calcination of metals Mersenne seems to hold a less

[53] On Mersenne's theory of knowledge see R.H. Popkin, *The History of Skepticism from Erasmus to Spinoza* (Berkeley and Los Angeles, 1979), pp. 129-40; and P. Dear (n. 48), pp. 23-79.

[54] Mersenne, *Questions Inouyes* (n. 52), p. 64.

[55] Mersenne, *Verité des Sciences* (n. 50), p. 56, and id., *Questions Inouyes,* (n. 52), p. 124.

[56] Mersenne, *Questions Théologiques* (n. 49), pp. 24-6.

sceptical position. Following Rey, he gives a clear corpuscular interpretation to tin's increase of weight after calcination.[57] In the section on occult qualities Mersenne maintains that those phenomena which have usually been explained as the effect of sympathy and antipathy are to be explained by means of atoms and chemical principles.[58]

Mersenne's adherence to the corpuscular theory of matter is also attested in *Harmonie Universelle*, where he deals with the causes of rarity and density. The explanation given by Mersenne is twofold: he resorts to both insensible particles and chemical substances. Tackling the problem at a more structural level, he sticks to a particulate theory, by stating that a body is denser if more particles are contained in a lesser space. When giving a more empirical explanation, Mersenne adopts the chemical principles theory, by stating that each principle brings about a specific set of qualities in bodies:

> Quelques-uns croyent que ces premieres causes ne sont autre chose que l'abondance d'esprit & de la quinte essence, & que plus un corps aura d'esprit, & plus il sera pesant, dur, dense; ce que l'on experimente au Caput mortuum des Chymistes, qui ne pese quasi rien, apres que le sel, le soufre, & le mercure en sont tirez. A quoy ils adjoustent que le sel, qui est la principale matiere du corps, leur donne la solidité, & qu'il les coagule, les fixe, & les congele, tant qu'il peut par sa vertu amalgamante, par laquelle les choses fluides et volatiles deviennent fixes et permanentes: par eemple, quand la pluye tombe, elle devient solide dans les vegetaux par le moyen de la terre; ce qui n'arrive pas quand elle tombe dans l'eau, dans la quelle le sel ne se reduit pas en acte; d'ou il s'ensuit que plus il y a de sel dans un corps, & plus il est dur, dense, pesant, & solide.[59]

From this passage we gather that Mersenne's interest in chemistry is not confined to marginal questions and cannot be dismissed as irrelevant to the understanding of his thought. I therefore reject Lenoble's distinction between Mersenne the author and Mersenne the correspondent. Mersenne's acceptance of chemical and corpuscular views, which is apparent in some of his works, is confirmed by his correspondence with Rey, Deschamps, Brun, Villiers, Stanihurst and van Helmont.[60]

[57] Ibid., pp. 11-5.

[58] Ibid., pp. 109-11.

[59] Mersenne, *Harmonie Universelle* (Paris, 1636), p. 203.

[60] Lenoble, *Mersenne* (n. 6), p. 70.

CHEMISTRY AND ATOMISM IN THE CIRCLE OF MERSENNE

The earliest chemical letter to be found in the correspondence of Mersenne is the one written in September 1625 from Rouen by Henry de Stanihurst, who was *curé* of Carentan, and connected with Lefèvre of Rouen, a less obscure correspondent of Mersenne.[61] The letter, containing answers to Mersenne's request for information on the transmutation of metals, makes some interesting statements on the composition of metals. Stanihurst maintains that the most perfect metals have more pure mercury in them and have their parts very closely united.[62] The explanation of the purification of mercury is here described along corpuscular and quantitative lines. In addition, it should be noted that the change of specific weight occupies a central position in Stanihurst's interpretation of the transmutative process:

> Il est manifeste aussy que ces impuretez tiennent les parties integrantes du mercure plus eloignées entre elles, et moins serrées, et par consequent la masse doit estre moins pondereuse soubs une mesme estendue locale. Or l'art peut purifier ledit mercure et le serrer de plus prez par la separation desdittes ordures de l'air et de l'eau elementaire; et par consequent ledit mercure pourra acquerir plus de poix pour peser autant soubs une moindre estenduë locale qu'elle faisoit soubs une plus grande, et pourra approcher de près au poix de l'or au lieu que le mercure ordinaire est esloigné du poix de l'or d'une sixieme.[63]

Mersenne's most famous chemical correspondent was Jean Rey, whose *Essays* (Bazas, 1630) have received much attention from historians of chemistry.[64] Rey's *Essays* deal with the increase of weight in metals upon calcination. His explanation has long been highlighted as an early example of the experimental and quantitative approach to chemistry. Indeed it is so. But in the *Essays* we also find a particulate theory of matter underpinning

[61] On Stanihurst, see Mersenne, *Correspondance*, i, 274. On Lefèvre, see ibid. pp. 324-5.

[62] Mersenne, *Correspondance*, i, pp. 280-2.

[63] Mersenne, *Correspondance*, i, pp. 282-3.

[64] Jean Rey (c.1582 - c.1645) was born at Le Bugue in Dordogne, M.D. at Montpellier, practised at Le Bogue. He was connected with several correspondents of Mersenne: Jean Brun, Deschamps, apothecary and physician of Bergerac, and Pierre Trichet of Bordeaux. A biographical account of Rey and a bibliography of his *Essays* is to be found in D. McKie's introduction to the reprint of the *Essays*, see *The Essays of Jean Rey* (London, 1951), pp. ix-xliv.

his explanation of the calcination of tin. Rey asserts that calcination of metals brings about the increase in weight, but only for tin and lead. This increase Rey explains as due to air, which has been rendered denser, heavier and adhesive by the long-continued heat of the furnace. Particles of air, he continues, mix with calx and stick to its more minute particles.[65] Rey believes that air, like all bodies (including homogenous ones), is composed of different kinds of particles, having different size and weight. In fluid bodies the lighter and subtler parts lie on the top.[66]

Rey's view of the three chemical principles is based on the corpuscular theory of matter. He states that the particles of salt, sulphur and mercury are "sensiblement different en tenuité & pesanteur."[67] Tackling the thorny question of why only tin and lead – but no other metal – show increase in weight, Rey restates the view that the action of fire on bodies brings about addition and subtraction of particles. He believes that tin and lead contain few volatile particles, while other bodies (including metals and minerals) contain an abundance of these particles, which are easily liberated by fire. The result is a decrease in weight of the body.[68]

Rey's associate, Theodore Deschamps stands out as an unambiguous advocate of atomism. His own version of the atomic philosophy is by no means mechanical, as in his view atoms are endowed with forces, both attractive and repulsive.[69] The first evidence of Deschamps's atomism is contained in his letter of 1635 to Pierre Trichet at Bordeaux, where he rejects the view that atoms are all spherical and maintains that motion is inherent to atoms.[70] From a letter to Mersenne written by Jean Brun on 22 April 1640, we learn that Deschamps adopted Hero's doctrine of *vacuola* disseminated among the atoms.[71] Deschamps's theory of matter is well illustrated in his letters to Mersenne. For him, atoms have shape, size, impenetrability, solidity and attractive power. They continuously move in

[65] Rey, *Essay* xvi, *The Essays*, (n. 64), p. 97 and *Essay* xxvi, ibid., p. 139.

[66] Essay xiii, ibid., pp. 77-8.

[67] Essay xii, ibid., p. 70.

[68] Essay xxvii, ibid., pp. 140-1.

[69] Théodore Deschamps (c. 1588 - ?) corresponded with Mersenne from 1640 to 1645. He studied in Leiden and practised medicine in Bergerac. See Mersenne, *Correspondance*, ix, p. 537.

[70] Mersenne, *Correspondance*, v, pp. 573-7.

[71] Mersenne, *Correspondance*, ix, pp. 276-7. On Hero's *Pneumatica*, see M. Boas, 'Hero's *Pneumatica*. A Study of its Transmission and Influence', *Isis* 40 (1949), 38-48.

the void and possess "une certaine naturelle amour pour se entrasprocher et joindre de certains costés, et certaine hayne pour se disjoindre et esloigner de certains autres, laquelle amour et hayne les fait estre en mouvement perpetuel."[72] Deschamps, who holds to a kinetic theory of heat, maintains that the atoms of light are the smallest and more active ones. Life, he argues, is the motion of the smallest atoms.[73]

Very little is known of Jean Brun, Master-Apothecary in Bergerac, who in 1629 wrote a letter to Rey, which apparently gave rise to the latter's *Essays*. In his letter, which was published with the *Essays,* Brun asked Rey why tin gained weight in calcination, while lead decreased.[74] From his letter to Mersenne of 22 April 1640, we gather that Brun, like his friends and correspondents Trichet, Rey and Deschamps, held a corpuscular theory of matter.[75]

Christophe de Villiers stands out among Mersenne's correspondents for expounding an articulate philosophy of nature, which is based on the notion of salt.[76] For Villiers, salt is responsible for all generation – of minerals, plants and animals. It is described as the material embodiment of formative virtues, of spirits and of seeds. He argues that the three chemical principles are different forms assumed by the same saline principle:

> Ces troys principes chymiques ne sont autre chose que le sel diversement dissout ou desseiché ou recuit par la chaleur naturelle du Soleil dans icelle plante, et puis anatomizé, par l'art.[77]

In a subsequent letter, dated 7 July 1640, Villiers explains the nature and properties of salt and criticises Mersenne's reduction of them to the mechanical attributes of atoms – a view which Villiers finds too metaphysical and remote from senses:

[72] Letter of 31 July 1640, Mersenne, *Correspondance*, ix, p. 539.

[73] Letter of December 1640, Mersenne, *Correspondance*, x, p. 369: "Maintenant touchant les animaux, il est evident que leur vie consiste en la chaleur, et la chaleur au mouvement des atomes plus menus, qui se meuvent dans les espaces des autres."

[74] 'Lettre du Sieur Brun qui a donné subject au present discours', Rey, *Essays* (n. 64), p. 12.

[75] Mersenne, *Correspondance*, ix, pp. 275-82.

[76] Villiers was physician at Sens. The role of salt as the main agent in nature had already been stressed by Joseph Duchesne, cf. N.E. Emerton, *The Scientific Reinterpretation of Form* (Ithaca and London, 1984), pp. 209-220. In a letter to Mersenne of 28 October 1640 Descartes expressed a critical view of Villiers's theory of matter, see AT, iii, pp. 211.

[77] Villiers to Mersenne, June 1640, Mersenne, *Correspondance*, ix, p. 426.

> Vous adjoustez que, si le sel est principe seul, il faudra necessairement que des atomes d'iceluy, les uns soient piquants pour le goust de sel, et les autres ronds pour le doux et huileux. Sur quoy je vous diray que cette meditation me plaist fort, mais estant trop metaphysique et hors de nos sens, et peut estre de nostre raison, qui ne peut concevoir qu'un atome ayt des points, il m'est dificile de l'admetre.[78]

Villiers's rejection of the mechanical philosophy did not entail the dismissal of the corpuscular theory as such. Villiers explains the generation of spirit from volatile salts in terms of atoms, or rather corpuscles. Though the term he uses is 'atom', it is apparent that he does not mean an indivisible particle of matter:

> Tellement que à parler vrayement, cet esprit ou [mercure] tant vanté, ne seroit autre chose sinon le sel de la plante, volatil ou plus subtil, qui s'unie estroitement, dans le tems de al fermentation, à son flegme ou menstrue et luy donne la force et vertue de la plante par une dissolution tres perfaicte de ses atomes plus subtils.[79]

However, the contrast of Villiers's chemical philosophy with Descartes's mechanism is apparent. Villiers knew the *Météores* of Descartes who, in turn, had received Villiers's letter to Mersenne of 4 October 1640. Here Villiers claims that Descartes's view of chemical principles shows little difference from his own. Villiers evidently refers to Descartes's letter to Mersenne dated 15 September, where he states that salt, water, oil and a subtle matter can be separated from the *caput mortuum*.[80] In fact, Descartes's view of salt as expressed in the *Météores* has nothing in common with Villiers's. When in the *Météores* Descartes spoke of salt, he meant just ordinary salt and described it along purely mechanical lines: "La saleure de la mer ne consiste qu'en ces plus grosses parties de son eau, que i'ay tantost dit ne pouvoir estre pliées comme les autres par l'action de la matiere subtile, ny mesme agitées sans l'entremise des plus petites."[81] In addition, Descartes rejected Villiers's notion of fixed spirit as entirely unintelligible.[82] The contrast between the two philosophies was pointed out in 1640 by Lazare Meyssonnier, physician of Lyons and correspondent of

[78] Mersenne, *Correspondance,* ix, pp. 470-1.

[79] Villiers to Mersenne June 1640, Mersenne, *Correspondance*, ix, p. 427.

[80] Mersenne, *Correspondance,* x, p. 105.

[81] Descartes, *Météores*, discours troisiesme, A.T., vi, p. 249.

[82] Descartes to Mersenne 30 July 1640. Mersenne, *Correspondance,* ix, pp. 518-9.

Mersenne.[83] Meyssonnier saw Villiers as a follower "de la philosophie desraisonable de Paracelse, duquel les reveries ont esté quintessentiées inutilment en d'autres galimatias par Severinus Danus hors toute confirmation d'une experience sensible". Meyssonnier unambiguously supported the sound geometrical reasoning of Descartes.[84] Villiers himself became increasingly aware of the difference between his own view of salt and Descartes's: "Mais je me trouve different d'avec luy [Descartes] en ce qu'il ne dit pas que les meteores froids, comme la neige, gresle, etc., soient faits par le sel comme j'estime."[85] When dealing with generation, Villiers distances himself from Descartes's mechanism: "Mr des Cartes recourt aux figure et mouvement de la matiere, qui ne peuvent rien à la vie, sinon en partie et comme principe passif."[86]

VAN HELMONT'S CORRESPONDENCE WITH MERSENNE

From 1630 to 1631 van Helmont and Mersenne exchanged a number of letters covering a variety of subjects, including magic, philosophy and medicine. Unfortunately, only van Helmont's letters remain. Yet they enable us to some extent to guess what Mersenne's lost letters contained, namely questions on natural philosophy and medicine. One of the subjects was Gaffarel's *Curiositez Inouyes* (1629), a controversial work on talismans, which Mersenne unambiguously rejected as impious.[87] Mersenne was evidently eager to know van Helmont's opinion about Gaffarel. In a letter dated 26 September 1630 van Helmont sent his views to Mersenne. He

[83] Lazare Meyssonnier (1602-72) studied medicine at Montpellier and became *médecin du Roy* in 1642. See *Nouvelle Biographie Générale*, ed. Hoefer, 46 vols, (Paris 1862-66), s.v.

[84] Meyssonnier to Mersenne 31 May 1640, Mersenne, *Correspondance*, ix, pp.358-9.

[85] Villiers to Mersenne, end of October 1640, Mersenne, *Correspondance*, x, p. 198. See Descartes, *Météores*, discours vi and vii, AT, vi, pp. 291-324. Cfr. E. Gilson, *Etudes sur le rôle de la pensée médiévale dans la formation du système cartésien* (Paris, 1967^2), pp. 102-37.

[86] Villiers to Mersenne 9/10 December 1640, Mersenne, *Correspondance*, x, p. 309.

[87] On Jacques Gaffarel see R. Pintard, *Le Libertinage érudit dans la première moité du XVIIe siècle* (Geneva, 1983^2), pp. 187-90. In 1625 Gaffarel published *Abdita divinae Cabala mysteria*, containing attacks on Mersenne, who in turn responded by publishing a short booklet, bearing the title *De Gaffarello Judicio* (s.l., 1625). See Mersenne, *Correspondance*, i, pp. 303-6.

recognised Gaffarel's profound knowledge of Kabbalah, yet he could not refrain from stressing that in Gaffarel's book there was a host of superstitious and impious ideas, as for instance those on talismans. When he expressed his own view of magic, van Helmont claimed that the power of imagination was the source of natural magic – a statement which sounds like a vindication of his own *De Magnetica Vulnerum Curatione* (1621).[88]

Since most of Jean Baptiste van Helmont's extant letters to Mersenne deal extensively with the constitution of bodies, a digression on van Helmont's theory of mixture and generation is requisite. This necessarily involves an assessment of van Helmont's atomism. In 1890 Kurd Lasswitz argued that van Helmont held a corpuscular theory of matter, which – in his view – had a considerable impact in the development of the particulate theory of matter.[89] Lasswitz's interpretation of van Helmont as a corpuscular philosopher – which has recently been endorsed by Newman – is in fact based on a number of references to atoms to be found in various tracts included in the *Ortus Medicinae*.[90] Some chemical reactions, for instance the supposed transmutation of iron into copper and the production of glass, are explained in corpuscular terms. Both are explained in terms of addition and subtraction of particles.[91] The case of glass is however illuminating as it shows van Helmont's limited use of corpuscles. He was able to produce glass from sand and ashes and to recover sand in the same quantity. Van Helmont contends that glass is not a homogenous substance, but a mere aggregate of particles. It is generated by a mechanical apposition of parts, not by the formative power of *semina* – as are the homogeneous substances.[92]

In my opinion, the role of atoms in van Helmont's works is marginal. For him, the main agents in nature are spiritual, non-corporeal entities. The most relevant piece of evidence quoted by Lasswitz and Newman to support the

[88] J-B. van Helmont to Mersenne, 26 September 1630, Mersenne, *Correspondance*, ii, pp 530-40. In this letter van Helmont rejected the Paracelsian doctrine of signatures, claiming that the spirit contained in the seed produces the form of plants. On the doctrine of signatures, see M.L. Bianchi, *Signatura rerum. Segni, magia e conoscenza da Paracelso a Leibniz* (Rome, 1987). On van Helmont's *De Magnetica Vulnerum Curatione*, see Pagel, *van Helmont* (n. 41), pp. 8-11.

[89] Lasswitz, *Geschichte*, i, pp. 343-351.

[90] Newman, *Gehennical Fire*, pp. 110-4.

[91] Van Helmont, 'Potestas Medicaminum', §§ 37-8, *Ortus*, p. 479.

[92] Van Helmont, 'Terra', § 14, *Ortus*, p. 56.

interpretation of van Helmont as corpuscularian is to be found in the tract entitled 'Gas Aquae'. Here van Helmont explains the passage of water to vapour and to gas not as the generation of a new substance, but as "extenuatio, propter partium extraversionem."[93] Water, according to van Helmont, contains the *tria prima* which change their reciprocal disposition. On these grounds, Lasswitz and Newman come to the conclusion that van Helmont saw water as a complex corpuscle, containing the three principles in different spatial disposition. Lasswitz and Newman thought that van Helmont saw the production of gas as a physical change, namely, a change in the disposition of the three principles within the corpuscles of water.[94] In my view, van Helmont's explanation of the passage of water into vapour and gas does not effectively support the conclusion that van Helmont's theory of matter was corpuscular. A closer look at the latter's somewhat puzzling explanation of the phenomenon in question may help towards an understanding of his views on water. Van Helmont's statement that water contains the three principles seems to contradict his oft-repeated tenet that water is an element (the only real element), which, as a pure substance, cannot be decomposed into the *tria prima*.[95] One possible explanation of the passage of 'Gas Aquae' under scrutiny is that van Helmont's aim is to prove that there is no essential difference between vapour and gas. This explains why here he does not have recourse to ferments as specifying agents.[96] Van Helmont explains this process as one which is generating not a new substance but simply a change of the same water. This circumstance accounts for his adoption of a physical explanation. What is rather puzzling is van Helmont's identification of the different 'parts' of water with the *tria prima*. Van Helmont himself felt it necessary to justify this statement. He maintains that his explanation is to be understood as an analogy: "When I suppose this I intend to remedy the weakness of our intellect, like astronomers do with their eccentrics."[97] In my view, van Helmont's above-mentioned statement that water contains – one within the other – the three

[93] Van Helmont, 'Gas Aquae', § 10, *Ortus*, p. 75.

[94] Lasswitz, *Geschichte*, i, pp. 345-6, Newman, *Gehennical Fire*, pp. 112-3.

[95] "Nunquam autem in aqua fieri trium primorum separationem, multoque minus essentialem transmutationem ullam." ('Elementa', §§ 15-16, *Ortus*, p. 53.).

[96] This explanation is suggested by Hooykaas, 'Het Begrip', p. 170.

[97] "Haec suppono, pro ut Astronomi suos excentricos, ut intelligendi imbecillitati nostrae, eatur obviam." ('Gas Aquae', §§ 8-9, *Ortus*, p. 74.). "Quod autem quandoque elemento aquae sua tria tribuerim, id analogice locutum est." ('Tria Prima Chymicorum…', § 54, *Ortus*, p. 407).

principles does not mean that for him water has different particles in itself, even less that it is formed of complex corpuscles. He believes that water is the homogeneous, simple material substratum. It is informed by *semina* and ferments in a way which is by no means mechanical.[98] In addition, it must be stressed that throughout his work van Helmont maintained that gas is ultimately produced by a specific seed. The purely material change, that is the attenuation of water parts into atoms, is preliminary to a process that is qualitative, not mechanical.[99]

The notion of atoms is employed in van Helmont's theory of mixture and of generation. As remarked by Newman, Helmontian atoms are identical with the *minima naturalia*, i.e. the smallest particles into which a substance may be divided. There is little doubt that for van Helmont *minima naturalia* are actual physical units. It is also apparent that they have qualitative determinations, not mechanical properties. In order to assess the role of atoms/*minima* in van Helmont's natural philosophy, I start by examining the Helmontian theory of mixture. Having rejected the Aristotelian theory of elements, van Helmont concluded that the question "utrum elementa maneant in mixto" was groundless.[100] Following Severinus, van Helmont maintains that seeds (*semina*) are the formative principle whence all natural bodies originate. They operate not as physical agents, but according to the ferment they contain. The latter is the formative spiritual agent which informs water.[101] The purely mechanical juxtaposition of parts (*partium*

[98] "Seminibus quidem concessum est, ex aqua, sua fingere concreta, suamque tragoediam per formarum defluxum ad interitum ludere." ('Gas Aquae', § 44, *Ortus*, p. 80).

[99] "Seminalis enim concreti proprietas, quae in Gas perseverat, vi frigoris, & dierum maturitate moritur & in pristinam aquam Gas redit". ('Complexionum atque Mistionum Elementalium Figmentum', §§ 29, *Ortus*, p. 108). On van Helmont's concept of gas see Pagel, *van Helmont* (n. 41), pp. 61-3 and G. Giglioni, 'Per una storia del termine *Gas* da van Helmont a Lavoisier: costanza e variazione del significato', *Annali della Facoltà di Lettere e Filosofia dell'Università di Macerata* 25-26 (1992-3), 431-68, esp. 439-40.

[100] 'Aër', § 11, *Ortus*, pp. 63-4.

[101] "Duo igitur, nec plura, sunt corporum, & causarum corporalium prima initia. Elementum aquae nimirum, sive initium ex quo: & fermentum, sive initium seminale, per quod, id est dispositivum, unde mox producitur semen, in materia."('Causae et initia naturalium' § 23, *Ortus*, pp. 35-6.). In van Helmont's philosophy there is a hierarchy of psycho-physical agents. Seeds are the most complex, they are matter-bound, whereas ferments are only by choice connected

appositio) does not bring about a real mixture. This is produced only by a spiritual agent.[102] Van Helmont's corpuscular views are mainly to be found in his oft-repeated statement that the reduction of bodies into their *minimae partes* is a 'pre-condition' for transmutation – which is ultimately a spiritual process. Van Helmont's limited use of the notion of corpuscles is also evident from his statement that penetration of masses occurs in nature. This, in my view, leaves little room for the corpuscular theory of matter as an *explanans*.[103]

We may now go back to van Helmont's letters to Mersenne, which impinge on the theory of matter. In a letter dated 15 January 1631 van Helmont answered Mersenne's question about the existence of void. For van Helmont, air (but not water) has pores and the existence of a vacuum in its interstices cannot be excluded.[104] He believed that the pores of air are filled with *magnale*, a substance which has an intermediate nature between body and non-body.[105] The entrance of *magnale* in the pores of air is responsible for its rarefaction. Van Helmont did not describe *magnale* in corpuscular terms (unlike the Cartesian *materia subtilis*). It is a semi-material substance, which can be identified with spirit, namely a principle of life and activity. The question of pores in bodies crops up in van Helmont's answer to Mersenne's question about the specific weights. Van Helmont, who accepts the existence of void, positively maintains that gold has pores – to which mercury can have access.[106] These corpuscular views by no means presuppose a mechanical theory of matter. We read in van Helmont's answer to Mersenne's question on the origin of hardness in bodies that *semina* are deemed to be the formative agents producing a qualitative change in bodies:

with matter. See 'Imago fermenti', §§ 12-13, *Ortus*, pp. 113-4, cf. Pagel, *van Helmont*, pp. 72-3.

[102] 'Progymnasmata Meteori', § 11, *Ortus*, p. 69.

[103] 'Progymnasmata Meteori', §§ 19-20, *Ortus*, p. 71.

[104] "Aqua enim vacuum non tolerat, ut neque sui compressionem, per aliquod movens comprimendo. Duntaxat inspissatione seminali comprimitur, per sui transmutationem formalem. Ex opposito autem, aer, sine vacuo subsistere nequit … ideoque sui compressionem, atque dilatationem tolerat." ('Progymnasmata Meteori', §§ 3-4, *Ortus*, p. 67).

[105] "Si l'air rarifié a du vide? Respondeo affirmative, et in poris aeris est ether sive magnale, quod est medium inter corpus et non." (van Helmont to Mersenne 15 January 1631, Mersenne, *Correspondance*, iii, p. 34.).

[106] Van Helmont to Mersenne, 30 January 1631, Mersenne, *Correspondance*, iii, p. 55.

> Quippe sali natura indurandi seu coagulandi in propriam et non alienam consistentiam, inest. Voco propriam, siquidem principiis singulis inest certa quaedam et non fallibilis, sed seminalis, scientia omnium dispositionum illius concreti seu individui, cuius sunt seminaria initia et Archeus sive causa efficiens interna (quam Aristoteles ignoravit; omnem causam efficientem externam indigetans, rustico ac plane mechanico intellectu) eadem principia concitavit ac disposuit et ad rerum agendarum fines scientiis propriis adornavit.[107]

As far as the chemical principles are concerned, in a letter to Mersenne of 15 January 1631 van Helmont maintained that they are "prima in synthesis et ultima in analysis". This somewhat positive view of the chemical doctrine of principles is mitigated by the statement that salt, sulphur and mercury cannot be obtained by means of fire. In his opinion, they may be obtained only by means of a more powerful solvent, like Paracelsus's *sal circulatum*.[108] Van Helmont's well-known (and influential) rejection of the chemical theory of principles is based on two arguments: 1. that water and *semina* are the ultimate constituents of natural bodies; 2. that they are not extracted, but produced *ex novo* by fire.[109] It is however apparent that van Helmont accorded mercury a special status. He saw it as a homogeneous and simple substance. The seed of mercury, as he put it, "is not a mortal one, not perishable, and does not obey the laws ruling sublunary bodies." All metals contain mercury, which is their kernel and the cause of firmness, while sulphur is their external and impure part.[110]

To conclude, the evident anti-materialistic point of view – which inspires much of van Helmont's rejection of the four elements and of the three chemical principles – is the reason why he imposed severe restrictions on

[107] Van Helmont to Mersenne, 11 January 1631, Mersenne, *Correspondance*, iii, p. 13.

[108] Van Helmont to Mersenne, 15 January 1631, Mersenne, *Correspondance*, iii, p. 31.

[109] See 'Complexionum atque Mistionum Elementalium Figmentum', § 10, *Ortus*, p. 105, 'Imago Fermenti Impregnat Massam Semine?', § 7, ibid., p. 112 and 'Tria Prima Chymicorum Principia', ibid., pp. 398-412. Pagel stressed the anti-materialistic overtones of van Helmont's criticism of the Paracelsian doctrine of principles, see Pagel, *Van Helmont* (n. 41), pp. 59-60.

[110] 'Tria Prima Chymicorum Principia', § 58, *Ortus*, p. 408. For a detailed analysis of van Helmont's views of mercury, see Newman, *Gehennical Fire*, pp. 146-51.

the corpuscular theory of matter. *Semina rerum* and ferments are the active principles on which all natural phenomena ultimately depend.[111]

THE CHEMICAL COSMOLOGY OF WILLIAM DAVIDSON

William Davidson (or Davisson) of Aberdeen was the first Professor of Chemistry in France, occupying this position at the Jardin du Roy from 23 July 1648 to 26 July 1651.[112] The lectures of the Scottish Professor achieved a rapid notoriety. They were attended by John Evelyn and Thomas Hobbes, while Sir Kenelm Digby was among Davidson's acquaintances.[113] His works contain an amalgam of Neoplatonism and Paracelsianism which bring about a rather complicated cosmology. Davidson's universe is organised after a hierarchical order: Mind, World-Soul, Nature and Matter. Following Plato's *Timaeus*, and by applying geometrical demonstrations, Davidson established a correspondence between elements, chemical principles and the regular solids.[114] In his rather complex cosmology atomism plays an important part. Davidson's first printed work (*Philosophia Pyrotechnica*, 1633-5) shows evidence of the fusion of the Neoplatonic-Paracelsian cosmology with a particulate theory of matter. His acceptance of the matter/form theory does not prevent him from describing the structure of natural bodies as composed of indivisible corpuscles, or atoms. These are activated by light, which is deemed a substance possessing both a corporeal and an incorporeal

[111] "Ego siquidem nudam lubens physicam aspicio ubique, non sane figuras; aut vires moventes in mathesi applico naturae". ('Imago Fermenti', § 7, *Ortus*, p. 112.) See also van Helmont, 'De lithiasi', iv, §§ 11-12, *Opuscula Medica Inaudita* (Amsterdam, 1648, first edn: Cologne, 1644), pp. 34-5.

[112] In 1651 Davidson moved to Poland, where he became physician to the Queen Maria Luisa Gonzaga. On Davidson see DSB; E.T. Hamy, 'William Davidson, Intendant du Jardin du Roy et Professeur de Chimie (1647-51), *Nouvelles Archives du Muséum d'Histoire Naturelle*, 3e série, 10 (1898), 1-38; Metzger, *Doctrines Chimiques*, pp. 45-51; J. Read, 'William Davidson, First Professor of Chemistry at the Jardin du Roi (1648)', *Archives Internationales d'Histoire des Sciences* 30 (1951), 660-66; id., 'William Davidson of Aberdeen. The First British Professor of Chemistry', *Ambix* 9 (1961), 70-101; Partington, iii, pp. 4-7. J-P. Brach 'Deux exemples de symbolisme géométrique dans des textes alchimiques du XVIIe siècle', in D. Kahn and S. Matton (eds.), *Alchimie. Art histoire et mythes* (n. 14), pp. 717-35.

[113] See Read (n. 112), 74; 77.

[114] See Brach (n. 112). See also R. Halleux's notes to Kepler's *L'Etrenne ou la neige sexangulaire* (Paris, 1975), p. 115.

nature.[115] Davidson accepts the four elements as well as the three principles, which he differentiates on account of the local disposition of parts, that is, their textures and the natures of the atoms of which they are composed:

> Ita ex atomis homogeneis facta sunt homogenea, ex heterogeneis, heterogenea. Ita corpora sunt vel contigua aut continua, ob harum atomorum inter se continuitatem, vel contiguitatem. Ex maxime continuis enim & heterogeneis facta sunt Sal, Ignis, & Mercurius. Ex minus continuis heterogeneis Aër, Aqua, & Sulphur. Ex minime continui & maxime homogeneis Terra rerum omnium mater. Ex quibus patet nihil esse reale in corporibus, praeter atomos, ex quarum diverso situ & primo charactere omnia corpora composita esse conspiciuntur.[116]

For Davidson, besides rarefaction and condensation, chemical experiments provide evidence for the existence of atoms, namely, of particles which remain unchanged in chemical reactions. The *reductio ad pristinum statum* is one of the chemical proofs used by Davidson to assert the existence of immutable corpuscles.[117] Though Davidson unambiguously stresses the role of Divine Providence in the origin and existence of the natural world, in *Philosophia Pyrotechnica* he often employs arguments taken from Lucretius.[118]

Davidson's commentary on Severinus's *Idea Medicinae* is much more than a mere explanation of his text. It is in fact a large work devoted to natural philosophy and chemistry, containing doctrines not to be found in Severinus's *Idea*.[119] The Neoplatonic cosmology, which is the foundation of his *Philosophia Pyrotechnica,* is reassessed in the commentary. The few elements of Aristotelian philosophy (which in the previous work were just a residue) disappear entirely. In Davidson's universe, which is arranged according to a hierarchy of entities, the seminal principles are given primary importance. Following Severinus, Davidson states that seeds (containing a spiritual substance) are endowed by God with knowledge (*scientia*).[120] As a

[115] W. Davidson, *Philosophia Pyrotechnica, seu Cursus Chymiatricus* (Paris, 1650^2), pp. 85-89; and 316.

[116] Ibid., p. 316.

[117] Ibid., pp. 317-31.

[118] Ibid., p. 326.

[119] W. Davidson, *Commentariorum in Sublimis Philosophi & Incomparabilis Viri Petri Severini Dani Ideam Medicinae Philosophicae Propediem proditorum Prodromus* (The Hague, 1660).

[120] Ibid., pp. 206-12.

spiritual plastic principles, *semina* have a higher status than the particles of matter, the latter having no activity and powers in themselves. The seminal principles are the source of activity and organise the confused mass of atoms (*confusa atomorum congeries*).[121] Davidson rejects the view that atoms are purely mathematical entities and produces a substantial amount of experimental evidence for their existence – which, as in *Philosophia Pyrotechnica*, is largely based on chemical experiments.[122]

PIERRE GASSENDI: ATOMS, *MOLECULAE* AND *SEMINA*

A key figure in seventeenth-century atomism, Pierre Gassendi is commonly described by historians as a champion of the new mechanical philosophy and as an opponent of Paracelsianism and of alchemy – mainly on the basis of his attacks on Robert Fludd.[123] It is the scope of the present section to reassess Gassendi's theory of matter and to investigate the relationship of his version of atomism with chemistry.

On closer analysis, Gassendi's atomism shows some substantial differences with the 'orthodox' mechanical philosophy. If we consider the latter as based on the notion of inert matter endowed with shape and size, and of motion as extrinsic to matter, it is manifest that Gassendi's view of matter is not purely mechanical. This interpretation of Gassendi's matter theory (which stresses its difference from the Cartesian one) may also facilitate the understanding of Gassendi's relationship to the chemical philosophy. Gassendi's notions of *semina* and spirits no longer appear as an illegitimate intrusion of Paracelsian ideas into the body of the 'sound' mechanical philosophy. Rather, these notions are to be understood as part of a theory of matter which does not dispense with forces, activities and powers. Chemistry and atomism are closely connected in Gassendi, whose

[121] Ibid., p. 79.

[122] Ibid., pp. 258-61 and 363. Davidson's experiments are taken mainly from Sennert and Sperling.

[123] The bibliography on Gassendi is huge. An updated bibliography is to be found in S. Murr (ed.), *Gassendi et l'Europe* (Paris, 1997), pp. 467-92. An exception to the standard view is Bloch, who showed the presence of chemical themes in Gassendi's philosophy, see O. Bloch, *La Philosophie de Gassendi. Nominalisme, Matérialisme et Métaphysique* (The Hague, 1971), pp. 252-9. The interpretation of Gassendi's theory of matter as strictly mechanist is reiterated by M.J. Osler, *Divine Will and the Mechanical Philosophy. Gassendi and Descartes on Contingency and Necessity in the Created World* (Cambridge, 1994), pp. 196-8.

probabilistic epistemology stressed the importance of intermediate causes (i.e. those based on the sensible qualities). Gassendi rejected Descartes's reductionist programme, the view that the only valid explanations of natural phenomena are those based on the mechanical properties of particles. In addition, Gassendi explicitly rejected the geometrisation of physics (and Descartes's identification of matter with extension). For Gassendi, geometry would not enable us to understand the variety of natural phenomena, and cannot be employed in the investigation of the microstructure of bodies (atoms, molecules and textures).[124]

In Gassendi's philosophy (like in Epicurus's) atoms are endowed with motion, which is an internal principle of activity. In the early version of Gassendi's atomism (*Animadversiones*, 1649) the motion of atoms is conceived as intrinsic to atoms. Matter, according to Gassendi, is active:

> Neque enim absurdum est facere materiam actuosam; absurdum potius facere inertem; quoniam qui talem faciunt, et ex ipsa tamen fieri omnia volunt, dicere non possunt , unde ea, quae fiunt, suam efficiendi vim habeant.[125]

Though Gassendi states that motion (like atoms) ultimately depends on God, his stress is on the internal motion of atoms, which are endowed with weight. In the subsequent *Syntagma Philosophicum* (published posthumously in 1658) Gassendi modified his position on the origin of motion, by rejecting the view that atoms have an internal principle of motion.[126] The question of the origin of motion, however, is not entirely decided by Gassendi. In the *Syntagma Philosophicum*, gravity (which is conceived as *propensio ad motum*) is still a property of atoms.[127] It is aparent that Gassendi could not easily eliminate the activity of matter from his natural philosophy. The ultimate source of motion in natural bodies (*res*

[124] *Syntagma Philosophicum*, in *Petri Gassendi Diniensis... Opera Omnia* 4 vols (Lyons, 1658) [hereafter *Opera Omnia*], i, p. 265b. Cf. T. Gregory, *Scetticismo ed empirismo. Studio su Gassendi* (Bari, 1961), pp. 157-9.

[125] Gassendi, *Syntagma Philosophiae Epicuri*, in *Animadvertiones in Decimum Librum Diogenis Laertii* (Lyons, 1649), repr. in *Opera Omnia*, iii, p. 19b. "Videlicet supponens motum (quem Democritus non negabat) convenire Atomis, absurdum censuit vim specialem ipsis non attribuere, qua talis motus cieretur: huiusmodi autem est gravitas, seu pondus, impulsiove, ac impetus, qua agi quicquid movetur, constat." Ibid., p. 201.

[126] 'Dicimus deinde explodendum esse, quod Atomi a seipsis habeant vim motricem, seu impetum.' *Syntagma Philosophicum*, *Opera Omnia*, i, p. 280.

[127] Ibid., p. 273.

concretae) is the motion of atoms. Atoms are endowed with gravity and have a natural tendency to disentangle in the bodies they compose. As Bloch and Messeri pointed out, the notion of *flos materiae* (the more subtle and active atoms) is often invoked by Gassendi in his physiology.[128] In all natural bodies, even in the most solid ones, the constituent particles are never at rest. The continuous motion is possible because in natural bodies are contained small interstitial *vacua*.[129]

On the whole, it is apparent that Gassendi's natural philosophy presupposes the notion of active matter. Gassendi clarifies this view by means of an analogy. Nature, he states, is to be compared to an army, not to a building, and God's relationship to the physical world is to be conceived as that of an emperor, not as that of a workman:

> Deinde igitur non sunt Naturae opera comparanda cum iis, quorum tota materia est iners, ac veluti mortua; sed cum iis, in quibus nisi omnes materiae partes, at aliqua saltem expers omnis motus, actionisque non est. Si comparatio igitur iuvat, confer opus naturae non cum domo, sed cum exercitu; materiam non cum lapidibus, sed cum militibus; agens non cum fabro, sed cum Imperatore.[130]

In Gassendi's natural philosophy the concept of molecule plays a central part.[131] Molecules are insensible corpuscles made up of different kinds of atoms.[132] Natural bodies are not always resolved into their constituent atoms, but are often analysed into molecules, or first concretions of indivisible atoms.[133] In addition, the concept of molecule provides the *explanans* for a variety of phenomena (mainly chemical and biological ones). It is a notion which helps bridge the gap between sensible qualities and atoms. The movements and powers of molecules originate from atoms. While atoms

[128] See Bloch (n. 123), pp. 268; and M. Messeri, *Causa e Spiegazione. La Fisica di Pierre Gassendi* (Milan, 1985), pp. 102-3.

[129] *Syntagma Philosophicum*, *Opera Omnia*, i, 277a.

[130] *Syntagma Philosophicum*, *Opera Omnia*, i, p. 336a.

[131] For Gassendi's notion of molecule, see Bloch (n. 123), pp. 252-9; Messeri (n. 128), pp. 109-12; and H. Kubbinga, 'La théorie moléculaire chez Gassendi', in *Quadricentenaire de la Naissance de Gassendi 1592-1992. Actes du Colloque International Pierre Gassendi. Digne-les-Bains, 18-21 Mai 1992*, 2 vols. (Digne, 1994), ii, pp. 283-302.

[132] "Heinc ex atomis conformari primum moleculas quasdam inter se diversas, quae sint semina rerum diversarum." *Syntagma Philosophicum, Opera Omnia*, i, p. 282b.

[133] *Philosophiae Epicuri Syntagma, Opera Omnia*, iii, p. 25b.

have all the same speed, and thus their motions can hardly account for natural phenomena, molecules are endowed with specific motions. Thus molecules – on the grounds of their different powers and textures – can explain natural phenomena better than the ultimate units of matter.[134]

Gassendi's molecules differ from Anaxagoras's *homoeomeriae* for three main reasons: 1. he rejects Anaxagoras's view that everything is made of everything (*omnia ex omnibus*), as different atoms or molecules have different powers in the generation of natural bodies; 2. unlike *homoeomeriae*, molecules can be resolved into atoms; 3. *homoeomeriae* differ according to their qualities, but Gassendi's molecules differ because of their textures.[135]

The notion of molecules is crucial for the understanding of Gassendi's views of chemistry. Gassendi is aware that mixed bodies are not always resolved into their constituent atoms, but are often analysed into molecules, i.e. compound corpuscles, which can hardly be decomposed. This is the case with the so-called chemical principles. Gassendi seems to maintain that the three chemical principles, as well as fire and water, are composed of specific molecules: "Huiusmodi moleculas esse quasi proxima, immediataque principia ignis, aquae & rerum magis simplicium, cuiusmodi Chymicorum quoque elementa, Sal, Sulphur, Mercurius & similia dici."[136] However, it would be misleading to infer that Gassendi adopted the chemical theory of principles. He did not deny that chemical analysis yields the substances chemists called principles. What he rejected (with theoretical, not experimental, arguments) was their claim that the three (or five) principles were simple and ultimate constituents of bodies. For Gassendi, the chemical principles can be further decomposed into their seeds, and ultimately into atoms ("in sua semina et postremum in Atomos").[137] *Semen* is here almost a synonym of *molecula*. As is apparent from Gassendi's definition of molecule, the notion of *semen* (which will be dealt with in the following pages) and that of *molecula* are equivalent in so far as they are both clusters of atoms.[138]

Gassendi's investigations of the properties of salts is based on their molecular structure, which is in turn determined by the *textura atomorum*. The reason why salt dissolves in water is the correspondence of its

[134] *Philosophiae Epicuri Syntagma, Opera Omnia*, iii, p. 20a.
[135] *Syntagma Philosophicum, Opera Omnia*, i, pp. 241ab and 472a.
[136] *Syntagma Philosophicum, Opera Omnia*, i, p. 472a.
[137] *Syntagma Philosophicum, Opera Omnia,* i, p. 245b.
[138] Cf. Bloch (n. 123), pp. 258-9.

molecules (cubes) with the empty interstices between the corpuscles of water. The solution becomes saturated when there are no more cubical *vacuola*. The small empty pores in the textures of bodies can have different forms. Therefore alum (whose corpuscles are octahedral) can be dissolved when salt cannot. The texture of saline corpuscles explains their actions on sense organs and their power of penetrating the most compact bodies. Likewise, solvents are deemed to dissolve different metals since the form of their corpuscles enables them to insinuate themselves into the pores of metals.[139] Unlike the Paracelsians, Gassendi explains the physical properties of metals, that is, their different specific weights, ductility and fusibility, by referring them not to mercury, salt and sulphur, but to their invisible *texturae atomorum*.[140] Gassendi's atoms are endowed with forms which are often purely arbitrary (*hamuli*, *ansulae*, etc.). It is however apparent (for example in the case with salts, gems and stones) that Gassendi also tried to deduce the invisible forms of atoms from the geometric forms of their crystals.[141] The transduction, that is the extensions of properties from the visible world to the microstructure of bodies, was indeed fairly common among seventeenth-century corpuscular philosophers and was ultimately based on the notion of nature's uniformity. Compared with Descartes, Gassendi made a rather limited use of transduction. The choice of the model in the visible world and the justification of transduction are more problematic for a philosopher who, like Gassendi, adopted a voluntaristic concept of God. Gassendi's voluntarism made the principle of nature's uniformity less effective than it was for other mechanical philosophers.[142]

Together with the mechanical explanations, Gassendi's *Physica* contains descriptive investigations of the physical and chemical properties of a number of substances. This is apparent in the section on salts. Gassendi distinguishes salts extracted from plants, which are fixed, from those of animals (*sal urinae*), which are volatile.[143] It is interesting to note that here Gassendi does not rule out the chemists' theory that salts are the cause of

[139] *Syntagma Philosophicum, Opera Omnia*, ii, p. 39a.

[140] *Syntagma Philosophicum, Opera Omnia*, ii, pp. 136a-b.

[141] *Syntagma Philosophicum, Opera Omnia*, ii, p. 114b. Emerton (n. 76), pp. 133-53.

[142] *Syntagma Philosophicum, Opera Omnia*, ii, 36b-37a. Cf. A.E. Shapiro, *Fits, Passions, and Paroxysms. Physics, Method, and Chemistry and Newton's Theories of Colored Bodies and Fits of Easy Reflections* (Cambridge, 1993), pp. 40-8.

[143] *Syntagma Philosophicum, Opera Omnia*, ii, p. 37b.

taste.[144] The section devoted to nitre, its extraction and generation (from subterranean effluvia) and the sections on alum and on antimony (whose curative power Gassendi, along with the Paracelsians, extols) confirm the relevance of Gassendi's descriptive and empirical approach to the study of nature – which marks the difference from Descartes's reductionist programme.[145]

Far from being the residue of an imperfect process of mechanisation of nature, the concept of *semina rerum* plays a prominent part in Gassendi's natural philosophy. As we have seen, Gassendi conceives *semina rerum* (like molecules) as compound corpuscles. Like molecules, seeds are endowed with activity – which ultimately originates from their component atoms. *Semina rerum* are not ordinary molecules. If all molecules have some inherent degree of activity, *semina* are endowed with formative power and 'programme', being responsible for the generation of minerals, plants and animals.

Gassendi ruled out both the view that stones originated from the beginning and the view explaining their origin by means of the two active qualities (heat and cold) transforming earth and water. The latter theory, which was supported, among others, by Agricola,[146] is rejected by Gassendi with experimental arguments.[147] Gassendi's claim is that stones are generated by a *vis lapidifica* or *virtus seminalis* implanted in the earth. This formative agent, which Gassendi calls *spiritus elaborator,* disposes the particles of matter into some regular pattern, which can be observed in a number of natural bodies, especially in minerals, stones and gems. Unlike rennet (*coagulum*), seeds operate according to a pattern, as in the corn.[148] In

[144] *Syntagma Philosophicum, Opera Omnia*, i, p. 411a.

[145] *Syntagma Philosophicum, Opera Omnia*, ii, pp. 37b-38b and 41b-42a.

[146] Georg Bauer (Agricola), *De Ortu et Causis Subterraneorum* (Basle, 1546).

[147] *Syntagma Philosophicum, Opera Omnia*, ii, p. 114a: "Deinde non videntur etiam calor, & frigus necessaria agentia, a quibus lapides creentur. Neque enim qui lapides in fluviorum fundis gignuntur, ob calorem concrescunt; neque qui intra Animalia formantur, compinguntur a frigore: & neque aut stillicidia, quae lapidescunt, aut fontes, qui vertunt res diversas in lapides, id a calore, aut frigore habent, cum absque utrovis idem faciant."

[148] "... non videtur profecto posse tales lapides fieri ex massa indiscreta, & quam non pervadat spiritus quidam elaborator, a quo illa partium, particularumque tam regularis distributio, & minorum in conformandis maioribus compactio fit." *Syntagma Philosophicum, Opera Omnia*, ii, p. 114b. "Iam ergo Lapidificum germen in succo collecto intra receptaculum totam interius massam pervadens corpuscula ita coaptat, ut simul massam coagulet, figat, ipsique duo quaedam praestet, quae non

the section of *Physica* devoted to gems Gassendi is more explicit: even the colour of precious stones is to be referred to the seminal power.[149] The analogy with biological processes is often repeated by Gassendi, who evidently found mechanical models of little use in the explanation of complex and organised structures. He rather resorted to a plastic agent, which he called *spiritus*. He maintains that *spiritus* is not an immaterial substance, but a material, subtle substance, composed of very active corpuscles.[150]

In the section of *Physica* dealing with plants Gassendi maintains that seeds contain the soul of the plant, i.e. a spiritual, tenuous and active substance, which is ultimately responsible for the generation.[151] This active principle is material and is made of special atoms, created by God at the beginning. The visible seeds of plants contain saline corpuscles which contribute to the development of the seed and to the formation of the plant.[152] The section on the generation of animals (which addresses the *vexata quaestio* of spontaneous generation) is based on the same combination of corpuscular theories and Creationist views as we found in the section on plants. The chain of motion of atoms and of molecules is the ultimate physical cause of the generation.[153] Yet generation is not explained in purely mechanical terms. For Gassendi, the seed of animals, which is composed of atoms coming from all the body, is animated. It contains a soul

possunt lacti praestari a coagulo. Unum, quod ipsam interea dispescat in plureis massulas, easque seu aequaleis, seu inaequaleis uniformiter configuret; habet id vero quatenus est non coagulum modo, verum etiam semen; idemque agit in materia lapidea intra conceptaculum, quod in materia triticea vis seminalis intra vaginam: nempe ut ex hac multa grana consimilia discernuntur, unde spica componitur; ita ex illa multi lapilli consimiles, unde contexitur rupicula gemmea." *Syntagma Philosophicum, Opera Omnia*, ii, p. 117b.

[149] "Colorum autem varietas referenda videtur partim ad seminalem vim, quae contexturae corpuscula uti ad specialem configurationem; ita ad specialem colorationem attemperet; partim ad commistionem succorum, qui ex Terris, Succis concretis, Mineralibusque aliis quidpiam coloratum delibaverit." *Syntagma Philosophicum, Opera Omnia*, ii, p. 118a.

[150] *Syntagma Philosophicum, Opera Omnia*, i, pp. 277b; 334a and 386a.

[151] *Syntagma Philosophicum, Opera Omnia*, ii, p. 172a-b. Gassendi maintains that the vegetative principle of plants (which is material) can be named *anima* only "ex analogia", ibid., ii, p. 145.

[152] *Syntagma Philosophicum, Opera Omnia*, ii, p. 171b. On salt as generative principle see Emerton (n. 76), pp. 209-226.

[153] *Syntagma Philosophicum, Opera Omnia*, ii, p. 275a.

bearing all the 'information' for the generation of the animal; since the seed is an epitome of body, the small soul therein contained is an epitome of the animal's soul.[154]

The so-called spontaneous generation was accepted by both Aristotelians and ancient atomists. For a Christian atomist like Gassendi it was a serious challenge. The production of living organisms from non-living matter could be considered a strong argument in favour of the particulate theory of matter, as the explanatory power of atomism was reinforced by showing that one uniform matter, made of corpuscles of different shape, size and motion, could generate living bodies. On the other hand, the materialistic implications of this view could not pass unnoticed. Gassendi's solution refers to invisible seminal principles created by God at the beginning.[155]

Gassendi is evidently aware of the risk of attributing a kind of intelligence (*cognitio*) to matter, which would mean reintroducing Epicurus's and Lucretius's materialism into atomism. Indeed Gassendi alleges that spirits have "notitiam sui operis", as it is apparent that they operate according to some programme. The nature of this knowledge is of course a thorny question for Gassendi. He has ruled out the existence of intermediate agents between God (and angels) and the physical world on the grounds that such an agent would reintroduce the *anima mundi*.[156] Gassendi's concern is to stress God's absolute power and His design in nature. In the section of *Syntagma* devoted to generation Gassendi has recourse to seminal corpuscles – a special kind of corpuscles originating not from the fortuitous aggregation of atoms, but by God's creative act. *Semina* (which contain spirits) – he stated – were formed by God at the beginning and are composed of special atoms. As Gassendi put it, God has formed "ex selectis atomis prima omnium rerum semina, ex quibus deinceps fieret per generationem propagatio rerum."[157] It is therefore apparent that Gassendi distinguishes two different kinds of atoms, i.e, 'ordinary' atoms and those

[154] "Intelligi deinde potest Animam, quae in semine, prout ipsa quoque defluxit ex omnibus partibus esse & consciam nutricationis, animationis, constitutionis singularum adeo, ut cum iam sit animae totius quasi epitome, agere idem pergat in materiam seminis, quae est ipsa quoque epitome totius corporis...", *Syntagma Philosophicum, Opera Omnia*, iii, 275b. Cf. J. Roger, *Les sciences de la vie dans la pensée française au XVIIIe siècle* (Paris, 1993^{2}), pp. 135-140.

[155] *Syntagma Philosophicum, Opera Omnia*, ii, p. 262b. See also ibid, pp. 170b-171a, on the spontaneous generation of plants.

[156] *Syntagma Philosophicum, Opera Omnia*, i, p. 334a.

[157] *Syntagma Philosophicum, Opera Omnia*, i, 280b.

endowed by God with a 'programme'. This is implicit in his sections on mineralogy and plants. When dealing with spontaneous generation Gassendi's position is more explicit; he denies that matter has an internal principle of organisation, maintaining that invisible *semina* (contained in the earth and water) were created at the beginning by God. A relevant consequence of this view is that the uniformity of matter (one of the glories of the mechanical philosophy) is lost in Gassendi, and, as will be apparent from the following chapters, in a number of corpuscular philosophers. For Gassendi matter is active, but has no internal principle of organisation – which ultimately depends on God.

GASSENDI ON ALCHEMY AND FLUDD

The standard view of Gassendi's view of alchemy is based on his attack on Robert Fludd, which was written on Mersenne's request.[158] Mersenne's (and Gassendi's) polemics with Fludd have been described as examples of the opposition between the quantitative and the alchemical, Hermetic and vitalistic view of nature.[159]

In order to understand Gassendi's objections to Fludd, some preliminary considerations are necessary. The *Exercitatio* (1630) against Fludd was published before Gassendi started his major work on Epicurus, and therefore it is not based on the atomic theory of matter. The issue in Gassendi's work against Fludd is not a mechanistic theory opposing the alchemical view of nature. What Gassendi criticises is Fludd's interpretation of alchemy in religious terms and the fusion of alchemy, cabbala and religion. Fludd's notion of *Anima Mundi* and his Biblical exegesis came also under attack in

[158] *Petri Gassendi Theologi Epistolica Exercitatio*...(Paris, 1630), repr. as *Examen Philosophiae Roberti Fluddi*, in *Opera Omnia*, iii, pp. 211-268. See Gassendi to Peiresc 2 December 1628, and also Mersenne, *Correspondance*, i, pp. 61-2; ii, pp. 86-7 and 132-41. Mersenne had also asked van Helmont's view of Fludd. Van Helmont's judgement of Fludd is contained in a letter dated 19 December 1630, see Mersenne, *Correspondance*, ii, p. 584.

[159] See F.A. Yates, *Giordano Bruno and the Hermetic Tradition* (London, 1964), pp. 434-40; L. Cafiero, 'Robert Fludd e la polemica con Gassendi', *Rivista Critica di Storia della Filosofia*, 19 (1964), 367-410, and ibid., 20 (1965), pp. 3-15, and J-C. Darmon, 'Quelques enjeux épistémologiques de la querelle entre Gassendi et Fludd', F. Greiner, *Aspects de la tradition alchimique au XVIIe siècle* (Paris and Milan, 1998), pp.63-84.

Gassendi's polemical tract.[160] Gassendi ruled out the Neoplatonic world spirit as the source of life. Fludd believed that *spiritus* was produced in the Sun by the Holy Spirit.

Fludd's interpretation of *Genesis* in alchemical terms – an interpretation which he borrowed from the pseudo-Paracelsian *Philosophia ad Athenienses* – is singled out as an example of the impious combination of the sacred text with natural philosophy.[161] The attempt to found alchemy on the Scriptures is, Gassendi thinks, harmful for both alchemy and religion. Though some notions of philosophy can be found in the Sacred Scriptures, their aim – Gassendi claims – is not to teach natural philosophy. Gassendi evidently rejects the view that alchemy and magic were part of the *prisca sapientia,* as well as the related synthesis of Moses's and Hermes's teachings.[162] Fludd's doctrine of the celestial harmony and his interpretation of the world soul as a divine entity are both criticised in Gassendi's answer. Fludd's notion of harmony presupposes, according to Gassendi, that the human intellect can establish the reasons for God's creative act. The latter is absolutely free and cannot be comprehended by us.[163] Fludd's doctrine of the world-soul could not escape Gassendi's judgement. Gassendi unambiguously dismisses Fludd's identification of *Anima Mundi* with God as impious. Indeed, Gassendi does not accept the very existence of the world-soul: "I hardly

[160] On Fludd, see W.H. Huffman, *Robert Fludd and the End of the Renaissance*, (London, 1988).

[161] Gassendi, *Examen*, *Opera Omnia*, iii, pp. 231b-232a. On Fludd's alchemical interpretation of the Scriptures see N.E. Emerton, 'Creation in the thought of J.B. van Helmont and Robert Fludd', in P. Rattansi and A. Clericuzio (eds.), *Alchemy and Chemistry in the 16th and 17th Centuries* (Dordrecht, 1994), pp. 85-101. For the *Philosophia ad Athenienses*, see Paracelsus, *Sämtliche Werke*, 1. Abteilung: *Medizinische, naturwissenschaftliche und philosophische Schriften*, Hrsg. von K. Sudhoff, vols 1-14 (München-Berlin, 1922-33), vol. 13, pp. 387-423.

[162] Gassendi rejects the doctrine of the ancient origins of alchemy, see *Examen*, *Opera Omnia*, iii, pp. 259a-b. Cf. A. Clericuzio, '*Alchemia Vetus et Vera*. Les théories sur l'origine de l'alchimie en Angleterre au XVIIe siècle', in D. Kahn and S. Matton (eds.), *Alchimie, art, histoire et mythes* (n. 14), pp. 737-48. On the *prisca sapientia* see D.P. Walker, *The Ancient Theology* (London, 1972).

[163] "Capite iam tertio tuetur Consonantias suas Macrocosmicas. Ut concedam vero licere, Pytagoreorum exemplo, harmoniam quandam inter membra Mundi praecipua constituere: hoc tamen videor posse dicere, non abesse a fabula quidquid de illa usquam somniatur. Etenim, ut Opifex Mundi rationem quandam habuit, cur hoc situ, hac mole, hac forma condiderit omnia; ita miselli homunculi videntur nimis temere rationem illam determinare." *Examen*, *Opera Omnia*, iii, p. 232a.

convince myself that some kind of power (*vis*) is diffused through the universe – a force, which, like the soul, joins the parts of the universe and produces some harmony among them."[164] In Gassendi's universe (as in Boyle's) God intervenes directly in nature, and there is no need of any intermediate agent, like the world-soul.

These criticisms of some central aspects of Fludd's philosophy do not necessitate the rejection of alchemy. Gassendi explicitly maintains that the alchemical *Opus* is possible – though he was convinced that no alchemist had achieved it so far.[165] In addition, Gassendi recognises the importance of alchemy both for the knowledge of nature and for medicine. Gassendi – who in this work holds a sceptical view of knowledge – gives alchemy a prominent position in the investigation of nature.[166] Since *semina rerum* are conceived the origin of all natural bodies, Gassendi maintains that the transmutations of metals can be achieved when the seed of gold is discovered.[167]

Gassendi's view of the *Opus* changes very little in the *Syntagma Philosophicum*, where he holds the view that metals are composed of sulphur and mercury. As he did in the *Examen*, Gassendi maintains that a seminal power (*seminalis vis*) is responsible for the generation of metals.[168] A preliminary to the transmutation is to 'open' gold and to extract *auri germen*. The transmutation of metals, like their generation, is interpreted as a biological process involving the action of a seminal formative agent. Such an agent is made of atoms, whose contexture brings about a molecule, which is endowed with a formative power.

It is apparent that for Gassendi alchemy is both possible and useful in natural philosophy. The interpretation of the alchemical *Opus* as a process

[164] *Examen, Opera Omnia*, iii, p. 236a.

[165] *Examen, Opera Omnia*, iii, p. 259a.

[166] "Hac de re vero quidquid sit, existimo negari non posse, quin duo quaedam valde utilia cognitionis genera debeantur Alchymiae. Unum est circa Naturam. Etsi enim intimos usque rerum naturalium penetrare non liceat, ut ipsarum essentias, discrimina, vireis, actiones, & agendi modos, proportionem item, atque contexturam cum radicali, & propria singulorum causa dignoscamus: veruntamen si quidpiam ex iis, quae res quasque interne componunt, cognoscere concedatur, illud profecto isti Arti acceptum referendum est. Haec enim est, quae Naturae librum sola evolvit, ac perscrutatur, cum ceterae omnes superficie tenus naturalia considerent." *Examen, Opera Omnia*, iii, p. 259a.

[167] *Examen, Opera Omnia*, iii, p. 259a.

[168] *Syntagma Philosophicum, Opera Omnia*, ii, p. 140a.

involving the seminal principle of gold, analogous to biological processes, confirms that Gassendi's atomism does not involve a strictly mechanical view of nature.

CONCLUSION

In early seventeenth-century France atomism and Paracelsianism were often associated in their opposition to Aristotelianism. Atomists made use of chemical experiments and theories; and in turn, corpuscular views of matter are to be found in a number of chemical and alchemical tracts. Atomists like Basso adopted the Neoplatonic (and Paracelsian) notion of spirit to account for the origin of the atoms' motion. Though Gassendi rejected Fludd's philosophy and the world spirit, he did not refrain from adopting relevant chemical views. Unlike Descartes, he did not conceive matter as passive, and introduced active principles in order to explain a number of natural phenomena. The notion of seminal principles, which was common among the Paracelsians (and played a relevant part in de Clave's works), receives unambiguous corpuscular interpretation in Gassendi, whose *semina* are corpuscles endowed by God with a specific programme. They have a double use, both explaining the generation of living organisms and also stressing (against the Epicureans) God's design in nature. As we shall see, Gassendi's theory of matter had a strong impact in seventeenth-century England and on Robert Boyle's philosophy.

CHAPTER 3

CHEMISTRY AND ATOMISM IN ENGLAND (1600 TO 1660)

THEORIES OF MATTER IN EARLY SEVENTEENTH-CENTURY ENGLAND: AN OUTLINE

As Charles Webster has shown, the spread of Paracelsianism in England before 1640 was not confined to the practical aspects of medicine. Medical and chemical theories, as well as cosmological and philosophical doctrines originating from Paracelsus and his followers, had wide circulation in the late sixteenth and early seventeenth century. This is confirmed by Richard Bostocke, John Dee, Sir Hugh Plat, Theodore Turquet de Mayerne, Robert Fludd and Edward Jorden, among others. Paracelsian ideas had a significant impact on natural philosophy, as demontrated by Nicholas Hill, Thomas Hariot, Francis Bacon and Walter Warner.[1]

Very little is known about Hill. He evidently embraced much of Giordano Bruno's cosmology, and in fact Mersenne attacked him as Bruno's disciple. He was also loosely associated with the Northumberland Circle.[2] Robert Kargon has described his *Philosophia Epicurea* (1601) as "a confused, self-

[1] C. Webster, 'Alchemical and Paracelsian Medicine', in C. Webster (ed.), *Medicine, Health and Mortality in the Sixteenth Century* (Cambridge, 1979), pp. 301-34. A different view was held by Kocher and by Debus, who suggested that before the 1640s the philosophical and medical theories of Paracelsus had little or no influence, the impact of Paracelsianism in England being confined to the practical aspects of medicine. According to Debus, chemical therapies were adopted, but the foundations of Galenic medicine were not shaken. See P.H. Kocher, 'Paracelsian Medicine in England: (ca. 1570-1600)', *Journal of the History of Medicine* 11 (1947), 451-80 and A.G. Debus, *The English Paracelsians* (London, 1965). As is apparent from the sources used in this chapter, Paracelsian medicine and philosophy were widespread in early seventeenth-century England.

[2] N. Hill, *Philosophia Epicurea, Democritiana, Theophrastica proposita simpliciter, non edocta* (Paris, 1601). On Hill see M. Mersenne, *Quaestiones celeberrimae in Genesim* (Paris, 1623), cols. 1837-8; G. McColley, 'Nicolaus Hill and the Philosophia Epicurea', *Annals of Science* 4 (1939), 390-405; J. Jacquot, 'Hariot, Hill, Warner and the New Philosophy', in J.W. Shirley (ed.), *Thomas Hariot, Renaissance Scientist* (Oxford, 1974), pp. 110-4 and H. Trevor-Roper, 'Nicholas Hill, the English Atomist', in *Catholics, Anglicans and Puritans. Seventeenth-century Essays* (London, 1989²), pp. 1-39.

contradictory *mélange* of the views of many thinkers."[3] What Kargon saw as a sign of confusion, was in fact a common pattern of early seventeenth-century atomism, namely, an amalgam of atomism and different philosophical traditions. Sir Henry Percy, ninth Earl of Northumberland, patron of science and natural philosopher, evidently found no contradiction in the co-existence of atomism and alchemy – the latter, he thought, was to be based on the corpuscular theory:

> The Doctrine of Generation and Corruption unfoldeth to our understanding the method general of all atomical combinations possible in homogeneal substances, together with the ways possible of generating of the same substance, as by semination, vegetation [...] with all the accidents and qualities rising from those generated substances, as hardness, softness, heaviness, lightness, tenacity, frangibility, fusibility, ductibility, sound, colour, taste, smell, etc. the application of which doctrine satisfieth the mind in the generation and corruption [...] which part of philosophy the practice of Alchemy does much further, and in itself is incredibly enlarged, being a mere mechanical broiling without this philosophical project.[4]

Sir Henry Percy's intellectual leanings were shared by his fellow prisoner in the Tower of London, Sir Walter Raleigh. While in the Tower, Percy and Raleigh pursued alchemical and chemical experiments. Raleigh's alchemical interest is well documented in his *History of the World* (1614), upon which he worked during his incarceration in the Tower.[5]

Following Lucretius and Bruno, Hill maintained that the universe is infinite and alive. Space and time are infinite and homogeneous.[6] Hill declares that vacuum must exist, or else a great many physical phenomena would be inconceivable: "effluxus & influxus, rarefactio, condensatio,

[3] R.H. Kargon, *Atomism in England: from Hariot to Newton* (Oxford, 1966), p. 15.

[4] H. Percy, *Advice to his Son*, ed. G.B. Harrison (London, 1920), p. 70, quoted in Kargon, *Atomism* (n. 3), p. 14. Percy's *Advice* was written about 1595. Percy's library bears evidence of his interests in alchemy and Paracelsianism, also in the works of Bruno and Della Porta, see G. Batho, 'The Library of the "Wizard Earl", Henry Percy ninth Earl of Northumberland, 1564-1632', *The Library*, 5th series, 15 (1960), 246-56.

[5] See P.M. Rattansi, 'Alchemy and Natural Magic in Raleigh's *History of the World*', *Ambix* 13 (1966), 122-38.

[6] On Bruno's reputation in England see H. Gatti, *The Renaissance Drama of Knowledge. Giordano Bruno in England* (London, 1989), pp 49-73; and S. Ricci, *La fortuna del pensiero di Giordano Bruno. 1600-1750* (Florence, 1990), pp. 56-63.

corruptio, generatio, gravitas, & levitas, motus, & alteratio, non immaginabilis absque vacuo."[7] Prime matter is formed of atoms, i.e., indivisible and unchangeable corpuscles. Atoms are infinite in number, have a variety of regular forms and are perpetually in motion.[8] There is no place in Hill's universe for Aristotelian forms, though Aristotelian elements do not disappear. Hill maintains that elements and chemical principles are formed of atoms. Elements are not changed in the *mixtum* which results from the juxtaposition of the *minima* (i.e. the elemental atoms).[9] A number of Paracelsian ideas are disseminated in Hill's *Philosophia Epicurea*. He describes man as a microcosm and believed in Paracelsus's elixir. He maintains that the process of generation originates from seeds containing "mechanical spirits", which possess formative power. Hill's *semina*, unlike those of Severinus, are units of matter, clusters of atoms, and spirit is a very subtle body.[10] Hill placed special emphasis on chemistry. He maintained that chemical experiments are an essential part of natural philosophy. Hill was also familiar with magnetism. He shared Gilbert's view that the earth was a magnet.[11]

While Hill's philosophy was mainly speculative, Hariot and Warner combined atomism with solid experimental science. An associate of Percy, Hariot worked on astronomy, mechanics, optics and mathematics.[12] Hariot's version of atomism was mechanical, unlike Hill's. He claimed that the shape, size and motion of atoms, and vacuum can explain all natural phenomena. Changes in weight are produced by interposing smaller atoms in the *vacua* between large ones.[13] His extant papers show that he used atomism mainly to account for physical phenomena, notably those related to optics. Hariot's atomism was refuted by Nathaniel Torporley as incompatible with the Christian religion. Torporley's anti-atomistic tract, entitled 'A Synopsis of the Controversie of Atoms' has been reproduced by Jacquot as an appendix to his 1952 article on Hariot (pp. 183-7). Written after 1622 (it quotes Bacon as Verulam), it contains information about Hariot's ideas of matter. From this document we learn that Hariot's atoms

[7] Hill, *Philosophia* (n. 2), prop. 47. See also prop. 67.

[8] Ibid., props. 1, 4, 15, 42, 58-9 and 76.

[9] Ibid., props. 4, 31, 56-7.

[10] Ibid., props. 1-2, 9, 35, 41.

[11] Ibid., props. 35 and 93.

[12] See J. Jacquot, 'Thomas Hariot's Reputation of Impiety', *Notes and Records of the Royal Society of London* 9 (1952), 164-87; J.W. Shirley (ed.), *Thomas Hariot* (n. 2), and id., *Thomas Hariot: a Biography* (Oxford, 1983).

[13] Cf. Kargon, *Atomism* (n. 3), p. 26.

are eternal, indivisible, unalterable. Empty spaces exist between them. Their weight is in proportion to their dimensions, the denser bodies being made of atoms touching each other on all sides. Homogeneous bodies are made of atoms of the same shape, or arranged in a regular pattern.[14]

Francis Bacon's theory of matter has been thoroughly investigated by Graham Rees and by Benedino Gemelli. The former has highlighted Bacon's debts to Paracelsianism, while the latter has provided us with an excellent study of the role of atomism in Bacon's thought.[15]

According to Bacon, matter exists in two states: spirits and tangible matter. Spirits are active; tangible matter is passive, cold and inert. Spirits operate on matter producing most of the processes observable in the terrestrial world.

Bacon's theory of matter was corpuscular; and, though he often referred to atoms, his atomism, as Gemelli has persuasively demonstrated, has little in common with Democritus's atomism.[16] Bacon never bestowed onto atoms mechanical properties. He described atoms as *semina rerum,* particles which are endowed with powers, i.e., with a variety of motions (including action at a distance). The *virtutes* of atoms enable them to assume all possible forms. The unity of prime matter is denied by Francis Bacon: "in the atom's body exist the elements of all bodies, and in the atom's motion and virtue exist the beginnings of all motions and virtues."[17] In *Historia densi et rari* (possibly written in 1622) Bacon rejects the existence of void. Rarity and density are explained by means of *plica materiae* (pliancy of matter) – a matter which can contract and expand in space.[18]

The difference beteween Bacon's and Democritus's atomism is stressed by Bacon in *Novum Organum* (ii, 234.34) along the following lines:

[14] See Jacquot, 'Hariot's Reputation' (n. 12). Original at BL, Birch MS 4458, ff. 6-8.

[15] G. Rees, 'Atomism and Subtlety in Francis Bacon', *Annals of Science* 37 (1980), 549-71; and B. Gemelli, *Aspetti dell'Atomismo Classico nella Filosofia di Francis Bacon e nel Seicento* (Florence, 1996).

[16] As Gemelli pointed out, Bacon often adopted Lucretius's terminology in the description of atoms. See Gemelli, *Aspetti* (n. 15), p. 153.

[17] Bacon, *De Principiis atque Originibus*, in F. Bacon, *Philosophical Studies c. 1611-1619, The Oxford Francis Bacon*, vol. vi (Oxford, 1996), pp. 202-3. According to Rees, *De Principiis* was written between 1610 and 1620, see ibid., p. xxix.

[18] "Non est vacuum in natura, nec congregatum, nec intermistum; inter terminos densi et rari est plica materiae, per quam se complicat et replicat absque vacuo.", *Historia Densi et Rari*, in J. Spedding, R. Ellis and D. Heath (eds.), *The Works of Francis Bacon*, 15 vols (Boston, 1860-4) [hereafter *Works*], ii, p. 303.

> Neque propterea res deducuntur ad Atomum, qui praesupponit vacuum et materiam non fluxam (quorum utrumque falsum est), sed ad particulas veras.

Bacon placed special emphasis on corpuscles having distinct properties, *particulae verae*, which he identified with *naturae simplices*.[19] The notion of schematism is also central to Bacon's theory of matter. The different qualities of bodies derive from their *schematismus*, which Bacon often defines as *textura*, which would mean the arrangement of their parts. However, this is not to be interpreted in mechanical terms, that is, as a purely spatial disposition of particles of inert matter. For Bacon, the *schematismus* changes according to the proportion of spirits (which, as we have seen, are endowed with activity and perception) and tangible matter. This is exemplified by what he wrote in *Historia densi et rari*:

> I must now pass on to the dilatations of bodies which are caused by the liberation of the spirits; that is, when they break out of the prisons of the grosser parts, which had confined them closely, and prevented them from dilating. For in bodies of a compact texture and strongly united in the bonds of their integral nature, the spirits do not perform their work of dilatation, before there be a solution of continuity in the grosser parts by corrosive and stimulating liquors with or without heat. And this is shown in the openings and dissolutions of metals.[20]

A crucial part in Bacon's theory of matter is played by the notion of spirits, that is, volatile substances contained in all natural bodies. Even though spirits are material, but very fine substances, composed of particles of different sizes, they have appetites, desires and impulses. Spirits, which have two components, air and fire, can be both inanimate and animate. Natural bodies owe their specificity to the spirits they contain. According to Bacon, inanimate objects contain inanimate spirits only; vegetables have both inanimate and vital spirits; animals have vital spirits only.

[19] On the notion of *naturae simplices*, see M. Fattori, 'Des Natures Simples chez Francis Bacon', *Recherches sur le XVIIe siècle* 5 (1982), pp. 67-74.

[20] See also Bacon's description of gold: "Gold is the only substance which hath nothing in it volatile, and yet melteth without difficulty. The melting sheweth that it is not jejune, or scarce in spirits. So that the fixing of it is not want of spirit to fly out, but the equal spreading of the tangible parts and the close coacervation of them: thereby they [the spirits] have the less appetite and no means at all to issue forth." *Sylva Sylvarum*, in F. Bacon, *Works*, ii, p. 600. According to Bacon, metals are generated *in matricibus terrae*. See *Sylva Sylvarum*, Bacon, *Works*, ii, pp. 449-50; and id., *Articuli Quaestionum circa Mineralia*, Bacon, *Works*, iii, p. 812.

The role of spirits is central to Bacon's physiology. As Rees put it, "the vital spirits regulate all vegetative functions of plants and animals. Organs responsible for these functions, for digestion, assimilation, etc., seem to act by *perceptio*, mere reaction to local stimuli, but these reactions are co-ordinated by the vital spirit. These functions flow from the spirit's airy-flamy constitution. The spirit has the softness of air to receive impressions and the vigour of fire to propagate its actions."[21]

Rees shows that Bacon's cosmology owes much to Paracelsianism, in particular to the works of Duchesne. Nonetheless, his theory of the 'quaternions' differs from the Paracelsians' theory of principles in many respects. First, salt was not regarded as having the same status as mercury and sulphur; second, they were not principles, but, as Rees puts it, each of them was a compound, material substance, a cluster of different "simple natures"; water, air and fire are not conceived as passive matrices, but as "active members of their quaternion".[22]

Though Walter Warner's writings on atomism were never published, they were known to a number of natural philosophers, including Hobbes, Wilkins and Ralph Bathurst.[23] Warner's views of matter may be summarised as follows: 1. It has real existence in itself; 2. it is homogeneous; 3. it is eternal; 4. it is impenetrable; 5. it is divided into atoms. 6. it is passive, i.e. in itself it cannot produce the phenomena of the physical world, but requires an external active principle, which he calls *vis radiativa*. The qualities of bodies derive from the various arrangements of atoms, having an infinite variety of

[21] G. Rees, 'Introduction' to F. Bacon, *Philosophical Studies* (n. 17), summarising what he previously published in *Ambix* in 1975 and 1977. Bacon's mature version of the 'quaternions' includes a third one, working as intermediate, see Rees's introduction to *Philosophical Studies*, p. lviii. On Bacon's spirit see also D.P. Walker, 'Francis Bacon and Spiritus', in A.G. Debus (ed.), *Science, Medicine and Society*, 2 vols. (New York, 1972), i, pp. 121-130. See also Gemelli, *Aspetti* (n 15), pp. 114-39. Gemelli has shown that Bacon's view of spirit is indebted to Telesio.

[22] Rees's scheme summarises Bacon's views as follows: Sulphur Quaternion: subterraneal sulphur; terrestrial oil and inflammable substances; terrestrial fire; sidereal fire. Mercury quaternion: subterraneal Mercury; water and non inflammable substances, air, ether. Intermediate quaternion: subterraneal salts, juices of animals and plants, attached animated and inanimate spirits, heaven of fixed stars. Bacon, *Philosophical Studies* (n. 21).

[23] See J. Jacquot, 'Hariot' (n. 2). On Warner (1570-1642/3) see J. Prins, *Walter Warner (ca. 1557-1643) and his Notes on Animal Organisms* (Utrecht, 1992). On Ralph Bathurst, see Frank, *Harvey,* pp. 68-9; 108-113.

forms.[24]

Warner's manuscripts have been thoroughly investigated by Jan Prins, who has published Warner's notes on animal physiology. Prins pointed out that Warner's physiological theories are based on chemistry. Warner frequently compares physiological processes and chemical reactions which occur in the laboratory. He regarded the human body as a chemical laboratory – a view which was to be developed by English physiologists in the 1650s and 1660s.[25]

In conclusion, it may be said that in England early seventeenth-century atomism was connected with chemical (often Paracelsian) theories and experiments. The central part played by the notion of *spiritus* in both Bacon and Warner demonstrates that some units of matter were conceived as corpuscles endowed with motion, powers and perception. It is therefore apparent that the alliance of atomism and chemistry which is the core of Boyle's research programme, and which became common in Restoration England, was initiated by a number of natural philosophers active in Britain in the early part of the seventeenth century. Though the roots of such an alliance may be found in the first decades of the century, its forms changed over time. One major factor of change was the publication of van Helmont's *Ortus Medicinae* in 1648 – which marked a turning point in the history of iatrochemistry. In addition, the theories of matter held by Descartes and Gassendi showed the potentialities (and the pitfalls) of a fully articulated corpuscular theory.

TWO 'ARISTOTELIAN ATOMISTS': DIGBY AND WHITE

Kenelm Digby was one of the most influential natural philosophers of mid-seventeenth century England. He advocated the corpuscular philosophy and promoted chemistry. His *Discours* on the 'powder of sympathy' achieved enormous popularity throughout the century, witness its many translations. Digby's theory of matter is contained mainly in *Two Treatises* (1644), which were written to prove the immortality of the soul. The first (and longer) tract, devoted to the nature of bodies, contains a detailed system of natural philosophy, combining the Aristotelian doctrine of elements with the Cartesian corpuscular theory of matter.[26] In Digby's natural philosophy

[24] BL, Additional MS 4394, ff. 384-403. See J. Jacquot, 'Hariot' (n. 2), pp. 118-20.

[25] See J. Prins, *Walter Warner* (n. 23), pp. 71-4.

[26] Sir Kenelm Digby, *Discours fait en une celebre assemblée touchant la Guerison de Plaies par la Poudre de Sympathie* (Paris, 1658), Engl. trans.: *A late*

the Aristotelian notions of form and of prime matter disappear. He claims that these concepts are part of the Aristotelian metaphysics, not physics.[27] Following Descartes, Digby maintains that matter is passive and local motion the universal agent in nature.[28] Though he calls the ultimate particles atoms, they are not the indivisible corpuscles endowed with size, shape and motion of classic atomism. Like Descartes, Digby maintains that there are no indivisible particles of matter and denies the existence of void. Digby's, unlike Descartes's theory of matter, is a development of the *minima naturalia* tradition, and does not eliminate sensible qualities from matter.[29] Digby rejected the view that chemical principles are the ultimate constituents of bodies by stating (like van Helmont) that they are produced (not extracted) by fire. He kept the four elements, which he reinterpreted in corpuscular terms. He asserts that the "elements must remaine pure in every compounded body in such extreme small partes as we use to call atomes."[30] The different combinations of the four elements bring about different classes of natural substances. If water is the basis and earth the dominating element, easily divisible bodies are formed (like mud, butter, honey); if the basis is water and fire predominates, inflammable bodies are formed.[31]

In Digby's philosophy the status of qualities is not univocally defined. Though he states that qualities ultimately depend on the disposition of parts of matter, in opposition to mechanical philosophers such as Galileo and Descartes, Digby maintains that sensible qualities are not the outcome of interaction between corpuscles and sense organs, but that they have real presence in bodies.[32] To the four Aristotelian qualities Digby adds density

Discourse Made in a Solemne Assembly, touching the Cure of Wounds by the Powder of Sympathy (London, 1658); and id., *Two Treatises. In the one of which, the Nature of Bodies, in the other, the Nature of Mans Soule*; *is looked into: in way of Discovery, of the Immortality of Reasonable Soules* (Paris, 1644). On Digby's theory of matter see Lasswitz, *Geschichte*, ii, pp. 188-207; Kargon, *Atomism* (n. 3), pp. 70-3; B.J.T. Dobbs, 'Studies in the natural philosophy of Sir Kenelm Digby', *Ambix* 18 (1971), 1-25; id., 20 (1973), 143-63; id., 21 (1974), 1-28; J. Henry, 'Atomism and Eschatology: Catholicism and Natural Philosophy in the Interregnum', *The British Journal for the History of Science* 15 (1982), 211-39.

[27] Digby, *Two Treatises* (n. 26), p. 344.

[28] Ibid., pp. 276-7.

[29] Ibid., p. 38. On the development of the notion of *minima naturalia*, see above, pp. 10-13.

[30] Ibid., p. 143, see also pp. 122, 135, 276.

[31] Ibid., pp. 131-2.

[32] Ibid., pp. 273, 344.

and rarity – the latter playing a central part in his philosophy. When bodies are rarefied, "the little atomes perpetually move up and downe in every space of the whole world, making their way through every body, will set on work the little partes to play their game."[33] For Digby, most phenomena in nature are produced by effluvia of atoms, including those commonly attributed to supernatural causes, as the healing performed by the 'powder of sympathy' – the very subject of his *Discours* of 1657. If rarity is responsible for the instability of the compounds, density is invoked by Digby to explain chemical affinity. Digby believes that bodies unite when they have a similar degree of density, as proven by the fact that liquid substances combine more easily.[34] Digby rejects Gassendi's explanation of the different solubility of salts in water, based on the correlation of the forms of corpuscles with those of *vacuola* in the water. Digby argues that the pores of the same body must have the same size and shape. So the pores of water all have the same form. Digby's own explanation of solubility is less simple than Gassendi's. It is based on the weight, the dryness and the density of different salts:

> The true reason of this effect is (as I conceive) that one salt maketh the water apt to receive another; for the lighter salt being incorporated with the water, maketh the water more proper to sticke unto an heavier, and by dividing the small partes of it to beare them up, that otherwise would have sunke in it. The truth and reason of which will appeare more plainle, if att every ioynt, we observe the particular steppes of every saltes solution. As soone as you put the first salt into the water, it falleth downe presently to the bottome of it; and as the water doth by its humidity pierce by degrees the little ioyntes of this salt, so the small partes of it are by little and little separated from one another, and united to partes of water. And so infusing more and more salt, this progresse will continue, until every part of water is incorporated with some parts of salt: and then, the water can no longer worke of itselfe but in conjunction to the salt with which it is united. After which, if more salt of the same kind be putt into the water; that water so impregnated, will not be able to divide it; because it hath not any so subtile partes left, as are able to enter between the ioyntes of a salt so closely compacted: but may be compared to that salt, as a thing of equall drynesse with it; and therefore is unapt to moysten and to pierce it. But if you put unto this compound of salt and water, another kind of salt that is of a stronger and a dryer nature then the former, and whose partes are more grossely united; then the first salt dissolved in the water, will be able to gett in betwixt the joyntes of the grosser salt,

[33] Ibid., p. 145.
[34] Ibid., p. 119.

> and will divide it into little partes; and will incorporate his already composed partes of salt and water, into a decompound of two saltes and water; untill all his partes be anew impregnated with the second grosser salt; as before, the pure water was with the first subtiler salt. And so it will proceed on, if proportionate bodies be ioyned, untill the dissolving composition do grow into a thicke body.[35]

Attraction is the main subject of the *Discours* on the powder of sympathy Digby delivered at Montpellier in 1657. Digby's conference dealt with the sympathetic cure of wounds – a cure to be performed at a distance – which was very popular in the seventeenth century.[36] The powder was applied not to the wound, but to the blood-stained bandage. Digby explained the cure in terms of effluvia of atoms – which are moved by attraction. Digby's version of the sympathetic cure is based on the view that homogeneous bodies have a strong power of attraction: "bodies which draw the atomes dispersed in the aire, attract unto themselves with a greater power and energy such as are of their own nature, then such as are heterogeneous, and of a strange nature."[37] Accordingly, the atoms of blood travelled back to their original source, carrying with them the atoms of the powder of sympathy, particularly spirit of vitriol.

In *A Discourse Concerning the Vegetation of Plants* (1660) Digby referred to the growth of plants as a fermentation, which he described as "the

[35] If the saturate solution is heated, more salt may be dissolved in it. For Digby, this "sheweth, that the reason of its giving over to dissolve, is for want of having the water divided into partes little enough to sticke unto more salt: which, as in this case the fire doth; so peradventure in the other, the acrimoniousnesse of the salt doth it." (Ibid., pp. 155-6). Digby evidently knew Gassendi's views on solubility already in 1642, the date of composition of *Two Treatises*. On Gassendi's position, see above, p. 66.

[36] Digby, *A late Discourse* (n. 26). The powder was supposed to cure the wound, no matter how distant it was from the weapon. With a few exceptions, nobody questioned the validity of the cure. As Jan Baptista van Helmont pointed out (*De Magnetica Vulnerum Curatione*, 1621), the discussion was not a *quaestio facti*, it was a *quaestio juris*, namely, whether the magnetic cure was licit or not. The origin of this cure is not clear. It became part of Paracelsian medicine via the ps-Paracelsian *Archidoxis Magica*. See S. Rattray (ed.), *Theatrum Sympatheticum Auctum* (Nuremberg, 1662); and W.D. Müller-Jahncke, 'Magische Medizin bei Paracelsus und den Paracelsisten: Die Waffensalbe', in P. Dilg and H. Rudolph (eds.), *Resultate und Desiderate der Paracelsus-Forschung* (Stuttgart, 1993), pp. 43-55. For J.B. van Helmont's version of this cure see, W. Pagel, *Joan Baptista van Helmont. Reformer of Science and Medicine* (Cambridge, 1982), pp. 8-13.

[37] Digby, *A late Discourse* (n. 26), p. 110.

intestine motion of atoms, producing a great dilatation of the body".[38] Like Sendivogius, Digby states that a nitrous salt, giving life to vegetables and animals, acts as a magnet, attracting the vital principle of the air, that is the universal spirit.[39]

Digby's combination of atomism with Aristotelianism was criticised by Alexander Ross, who could easily resort to Aristotle's arguments against atomism.[40] Ross simply reassessed the role of elements and of primary qualities and maintained that in the *mixtum* the forms of elements do not disappear, but are *refractae*. He rejected Digby's view that effluvia are streams of atoms by saying that they are just vapours or fumes. Finally, Ross maintained that physics is subordinate to metaphysics and the principles of natural philosophy are to be found in Aristotle's metaphysics.[41]

Digby's compromise of atomism and Aristotelianism found an advocate in Thomas White, whose *Institutionum Peripateticarum libri quinque* are admittedly based on Digby's *Two Treatises*.[42] According to White, the four elements are made up of small parts, which are not atoms, though they are very seldom divided.[43] Like Scaliger and Sennert, White explains the *mixtio* in corpuscular terms: he states that the *mixtio* occurs when the elements are reduced to their smallest parts.[44] Strictly speaking, for White there are no pure elements, as the particles of light can easily penetrate what we consider elements.[45] Though White does not give a precise definition of the particles of matter, he maintains that bodies are made of particles of different density, size and weight, having pores of different sizes and shape among them.[46] In order to confirm the existence of corpuscles, White has resort to chemistry,

[38] Digby, *A discourse Concerning the Vegetation of Plants. Spoken at Gresham College on the 23 of January 1660* (London, 1661), pp. 12-3.

[39] Ibid., pp. 61-4 and 70.

[40] A. Ross, *The Philosophicall Touch-stone: or Observations upon Sir Kenelm Digbie's Discourses of the Nature of Bodies, and of the Reasonable Soule* (London, 1645).

[41] Ibid., pp. 19, 27, 59-61.

[42] Th. White, *Institutionum Peripateticarum libri quinque* (London, 1646-7); Engl. tr.: *Peripateticall Institutions. In the Way of that Eminent Person and Excellent Philosopher Sir Kenelm Digby* (London, 1656). On White, see B.C. Southgate, 'Covetous to Truth'. *The Life and Work of Thomas White*, 1593-1676 (Dordrecht, 1993) and J. Henry, 'Atomism' (n. 26).

[43] White, *Peripateticall Institutions* (n. 42), p. 56.

[44] Ibid., p. 60.

[45] Ibid., p. 69.

[46] Ibid., pp. 63, 67.

notably to the process of fermentation, which in his view shows that the motion of the particles of fire brings about major changes in natural bodies.[47]

EPICUREAN ATOMISM: MARGARET CAVENDISH'S POEMS AND JOHN EVELYN'S LUCRETIUS

The revival of Epicureanism in England was mainly due to the English emigrés to France. The so-called Newcastle Circle played a prominent part in English atomism. Margaret Cavendish, Sir Charles Cavendish's sister-in-law, and John Evelyn were part of the Newcastle Circle.[48] Both of them (though in different ways) contributed to the diffusion of Epicurus's philosophy in England. In the 1650s Margaret published a number of works expounding atomism and a fairly articulate though somewhat inconsistent philosophy of nature. In 1653 she published two works based on atomism, *Poems and Fancies* and *Philosophical Fancies.* She shares the Epicurean theory of the origin of the worlds from the fortuitous concurrence of atoms.[49] All atoms are made of the same matter, which is eternal and has innate motion. They have the same quantity of matter and therefore the same weight.[50] Margaret Cavendish believed that the motions of atoms are directed by an internal principle of action which she styles sympathy. Atoms which sympathise unite forming different natural bodies.[51] Unlike other Epicureans, she does not rule out the existence of the elements, though she deems them corpuscles having different geometrical forms: square = earth; round = water; long and straight = air; sharp = fire.[52] Margaret's theory of life and sensation is entirely materialistic. Life is produced by the motion of fiery atoms, when round atoms prevail, death occurs.[53] Life is the result of the rapid motion of fiery atoms. Spirits, which are the thinnest parts of bodies, are responsible for sensation and reason. Margaret adopted a chemical analogy to describe the rational spirits: they are "like little sphericall bodies of Quicksilver, several ways placing themselves in several Figures, sometimes moving in measure, and in order and sometimes out of

[47] Ibid., p. 64.

[48] On Margaret Cavendish and the Newcastle Circle, see Kargon, *Atomism* (n. 3), pp. 68-76.

[49] M. Cavendish, *Poems and Fancies* (London, 1653), p. 5.

[50] Ibid., p. 8.

[51] Ibid., p. 9.

[52] Ibid., p. 6.

[53] Ibid., pp. 14-19.

order."[54]

Evelyn's translation of Lucretius's *De rerum natura* evidently originated in the context of growing interest in Epicurean philosophy among French intellectuals.[55] While in France, Evelyn became familiar with the works of major French philosophers (Descartes, Gassendi, Naudé and La Mothe le Vayer) and was linked with exiled English philosophers.[56] Digby's and Charleton's works were no less important for Evelyn's adoption of atomism.[57]

Evelyn was concerned about Lucretius's reputation of atheism and tried to absolve the Latin poet of some of the charges of impiety. Yet the choice of translating Lucretius's *De Rerum Natura* with a substantial commentary could be conceived as a tribute to Epicureanism. Evelyn justifies his choice by asserting that irreligious ideas are to be found in a variety of philosophical texts currently available, including Plato's, Aristotle's and those of the Stoics. Modelling himself on Gassendi, Evelyn set out to purge Lucretius's verses from the impious doctrines contained therein. He maintained that it was not Lucretius who was the first to deny providence and the power of God, but it was Leucippus.[58] However, in the verses composed by Evelyn's father-in-law, Richard Brown, and published in the preface to the translation, Lucretius is styled as "the Oracle of all that can be knowne".[59] Evelyn published only the *Essay on the First Book of Titus Lucretius Carus de Rerum Natura*, with extensive animadversions, and left the translation of books iii-vi unpublished. This fact has been explained by Hunter as the outcome of Evelyn's growing anxiety about the impious views in the poem.[60]

The existence of void is accepted by Evelyn, who maintained that "so

[54] *Philosophicall Fancies* (London, 1653), pp. 38-9, and id., p. 64. Margaret's recantation of Epicureanism was confined to the doctrine of the fortuitous concourse of atoms. See *The Philosophical and Physical Opinions* (London, 1655).

[55] See T.F. Mayo, *Epicurus in England* (Dallas, 1934), *passim*.

[56] See *The Diary of John Evelyn*, ed. E.S. de Beer, 6 vols. (Oxford, 1955), iii, pp. 20 and 41.

[57] On Digby's influence, see BL Evelyn MS 32 (ca. 1649), fols. 7^v and 13^r. Walter Charleton was the other English philosopher who influenced Evelyn. Cf. M. Hunter, *Science and the Shape of Orthodoxy* (Woodbridge, 1995), pp. 67-98. See also H. Jones, *The Epicurean Tradition* (London, 1989), pp. 203-5.

[58] J. Evelyn, *An Essay on the First Book of Titus Lucretius Carus de Rerum Natura* (London, 1656), p. 106.

[59] Ibid., p. 2.

[60] Hunter, *Science* (n. 57), pp. 87-92.

frequent is this inanity, that even the most solid concretes have no contexture without it".[61] In his view, elements and atoms are not mutually exclusive.[62] Evelyn, who in Paris attended Lefebvre's chemical courses, regarded fermentation as the cause of most natural phenomena.[63] In Evelyn's extant manuscript notes on chemistry (covering the period between 1646 and 1652) we find the doctrine of the five principles and the view that the spirit of the world is the universal agent contained in all individual substances.[64] Though mainly practical in orientation, Evelyn's chemical (and alchemical) notes show that that in 1646 he had already adopted the atomistic theory of matter. This is borne out by his statement that menstrua dissolve bodies "into indivisible atoms".[65]

SEMINAL ATOMS: HIGHMORE'S THEORY OF GENERATION

Digby's views of generation are criticised by Nathaniel Highmore in his *History of Generation*, which is based on the notion of seminal atoms.[66] Unlike Digby, Highmore unambiguously rejected the doctrine of elements and qualities, which he replaced with the atomistic theory of matter. In his view atoms are not to be conceived as units of inert matter, rather they are endowed with powers. The notion of form still plays a part in Highmore's theory of generation, albeit a marginal one.[67]

In Highmore's opinion, the generation of living bodies is produced not merely by the addition of atoms, but requires the action of some kind of formative agent.[68] Whereas the generation of plants is achieved by seminal principles, which have their own formative power in themselves, the

[61] Evelyn, *An Essay* (n. 58), p. 134.

[62] Ibid., pp. 153-5.

[63] Ibid., p. 154. Evelyn's attendance at chemistry courses in Paris is attested by *The Diary* (n. 56), ii, pp. 534, 565; iii, p. 49.

[64] BL Evelyn MS 61 not paginated, ch. iv and v. See F. Sherwood Taylor, 'The Chemical studies of John Evelyn', *Annals of Science* 8 (1952), 285-92. Evelyn quotes Sendivogius as his source of the spirit of the world doctrine.

[65] BL Evelyn MS 32 fol. 39^r. The notes are dated 1646.

[66] N. Highmore, *The History of Generation* (London, 1651). On Nathaniel Highmore (1613-1685), see DSB, J. Roger, *Les sciences de la vie dans la pensée française au XVIIIe siècle* (Paris, 1993, 1st edn: 1963), pp. 106-11; 134-5, and Frank, *Harvey*, *passim*.

[67] See Highmore, *History* (n. 66), p. 44. Cf. H.B. Adelmann, *Marcello Malpighi and the Evolution of Embryology* (Ithaca and New York, 1966), pp. 777-9.

[68] Highmore, *History* (n. 66), pp. 24-7.

generation of animals requires the intervention of a spiritual agent. The principle responsible for the formation of a foetus is an immaterial and spiritual agent.[69] In human beings, the formal principle is the immortal soul, which Highmore (like Sennert) believes is communicated by the parents.[70] By invoking an immaterial principle as the formative agent, Highmore does not imply that atoms are particles of inert, passive matter. Once the seminal atoms have been disposed into the right places by the soul, they fashion themselves. The soul provides the 'programme', and atoms provide matter and energy for the generation of the new individual.[71] Highmore maintains that seminal atoms are of two kinds: spiritual atoms and more material atoms – the former is the masculine seed, the latter feminine. The duty of spiritual atoms is "to actuate, to enliven and to act", that of material ones is "to fix and cement the spiritual atoms together".[72] According to Highmore, the different powers of seminal atoms are responsible for the sex and any resemblance to the parents.[73]

Highmore explains spontaneous generation as the "mutual juncture of such Atomes, which before lay scattered in the bowels of some other compound; and wanted nothing, but union to fashion them into such a frame and structure." The seminal atoms unite themselves and, with the help of heat, produce a creature different from the original.[74] The process of growth is explained in chemical and corpuscular terms, a view which a number of English physiologists adopted in the mid-seventeenth century. The nourishment of the parts is effected by a tincture, which in turn is extracted from food by means of purification – to wit a series of distillations, concoctions and circulation. In this way atoms which are 'cognate' to the parts are selected and assimilated.[75]

Highmore's atomism is by no means mechanical. His atoms are not particles of inert matter, they have motion and powers. In addition, they are differentiated not by their sizes and geometrical forms, but by their different degrees of activity.

[69] Ibid., pp. 27, 53.

[70] Ibid., pp. 28-9.

[71] "The several Atomes fall to their respective places: the soul playing the skilful Workman (not laying brick where should be mortar) reposing every Atome in its proper place, that very same which it should have held in the body, from whence it was separated." (Ibid., pp. 85-6). See also p. 111.

[72] Ibid., p. 89.

[73] Ibid., pp. 91-3.

[74] Ibid., pp. 58-60.

[75] Ibid., pp. 40-1.

HELMONTIAN IATROCHEMISTRY AND ATOMISM IN THE 1650s

The 1650s saw a large spread of iatrochemical ideas in England, originating from Paracelsus and his followers and also from van Helmont (*Opuscula Medica,* 1644 and *Ortus Medicinae,* 1648). English iatrochemists often combined chemical and corpuscular views – an eclectic attitude which survived throughout the seventeenth century. Compared with iatrochemistry, Cartesian mechanical physiology had little impact in England. Even those who explicitly adopted Descartes's theory of matter, for instance Henry Power, had frequent recourse to notions taken from chemistry, mainly in the explanation of physiological phenomena.[76] From the 1650s and throughout the subsequent decade, though Paracelsus's works were read and translated, Helmontianism became a leading force in English chemistry and medicine and had a strong impact on Robert Boyle.

The diffusion of Helmontianism in England took place mainly via the Hartlib Circle. Early references to van Helmont, some dating 1644 (the year of publication of *Opuscula Medica*), stress van Helmont's opposition to Aristotelianism and to Galenism.[77] Among the advocates of van Helmont may be found William Rand (who also translated Gassendi's life of Peiresc into English), Robert Boyle, and George Starkey (Philalethes). Starkey played a major part in the early diffusion of Helmontianism and gave a corpuscular interpretation to van Helmont's chemical theories. As the author of the alchemical tracts published under the pseudonym of Philalethes, Starkey espoused an articulated corpuscular theory of matter, which has its sources in the *minima naturalia* tradition. As Newman has shown, Philalethes's alchemy was both corpuscular and vitalistic. Philalethes's elements are *minima* of a different size. Along with van Helmont, Philalethes stated that all mixed bodies originate from water and from *semina*, the latter

[76] On Descartes's physiology see A. Georges-Berthier, 'Le mécanisme cartésien et la physiologie au XVIIe siècle', *Isis* 2 (1914), 37-89; R.B. Carter, *Descartes' Medical Philosophy. The Organic Solution to the Mind-Body Problem* (Baltimore and London, 1983); A. Bitbol-Hespériès, *Le Principe de vie chez Descartes* (Paris, 1990) and F. Duchesneau, *Les modèles du vivant de Descartes à Leibniz* (Paris, 1998). On Henry Power see C. Webster, 'Henry Power's Experimental Philosophy', *Ambix* 14 (1967), 150-78.

[77] J.B. van Helmont, *Opuscula Medica Inaudita* (Cologne, 1644). See H. Appelius to S. Hartlib, 13 August 1644 HP 45/1/12. Early references to van Helmont are to be found in Sir Cheney Culpeper's letters of 1645 see HP 13/9/6A-B. On van Helmont and the Hartlib Circle see A. Clericuzio, 'Helmontianism and the Hartlib Circle', forthcoming in S. Mandelbrote (ed.), *The Hartlib Papers: a Universal Correspondency*, forthcoming.

having the power to produce fermentation. The chemical *mixtio* occurs when substances are reduced to their smallest parts, all of the same size. As Newman has pointed out, Philalethes had a theory of complex corpuscles, which was central to Boyle's and Newton's chemistry. Philalethes believes that corpuscles have different layers corresponding to the different chemical principles. Metals are conceived as complex corpuscles, with an internal kernel, which is a compact mercurial substance, linked to its 'essential sulphur', and an external porous shell which Philalethes calls 'external sulphur'.[78]

In the early 1650s three Helmontians emerged as the most radical opponents of traditional medical learning: they were John French, Noah Biggs and John Webster. Following Paracelsus and van Helmont, French, who was in touch with the Hartlib Circle and in particular with Robert Child, unambiguously rejected the Galenic doctrine of humours and advocated the chemical analysis of blood. He claimed that by means of distillation the chemical physician could analyse blood into its components: spirit, oil, water and salt.[79] French laid special emphasis on the extraction of spirits from blood and urine, which he, like van Helmont, interpreted as substances with strong therapeutic virtues.[80]

Noah Biggs's *Mataeotechnia Medicinae* (1651) vigorously amplified van Helmont's criticisms of traditional medicine, arguing that no real progress had been made in this discipline since antiquity. He claimed that physicians, by following the doctrines of Galen and other heathen philosophers, had relied upon fallacious reasoning, so had confined their investigations to the mere surface of things; whereas Helmontian chemistry relied on divine revelation, and was therefore capable of penetrating the "hidden things of nature" and "the maturation of seminall vertues".[81] God endowed natural

[78] See Newman, *Gehennical Fire*, pp. 141-164.

[79] See J. French, *The Art of Distillation* (London, 1653, 1st edn: 1651), p. 89. For French, who served as physician to the parliamentary army, see Webster, *Instauration*, p. 279. French was acquainted with Johann Brun, who was interested in Helmontian chemistry since 1648.

[80] French, *The Art of Distillation* (n. 79), pp. 92-4.

[81] N. Biggs, *Mataeotechnia Medicinae Praxeos. The Vanity of the Craft of Physick* (London, 1651), pp. 50, 57-8, 112. Biggs's work has been discussed by A.G. Debus, 'Paracelsian Medicine: Noah Biggs and the problem of Medical Reform', in A.G Debus (ed.), *Medicine in Seventeenth Century England* (Berkeley, 1984), pp. 33-48. Very little is known of Noah Biggs, even his identity is obscure. Cook suggests that Noah is the pseudonym of either Thomas Biggs or of his son Henry Biggs. See H.J. Cook, *The Decline of the Old Medical Regime in Stuart*

bodies with a "lumen quoddam vitale", whereby they act one upon the other. This *lumen* is the vital spirit, the source of "all activity, capacity and power".[82] Biggs's work is relevant because it is one of the first chemical treatises published in England to oppose the Paracelsian doctrine of principles by using van Helmont's arguments. Biggs states that the three principles "are new created things", produced by fire.[83] William Johnson's immediate reply (1651) to Biggs's *Mataeotechnia* shows that what was at issue was not the acceptance of Helmontian chemical medicine as such: rather the arguments were about the interpretation of van Helmont's texts and the use of his works in the disputes about established medical learning.[84]

John Webster's controversial *Academiarum Examen* mentions van Helmont only in a tangential way. In the chapter of *Academiarum Examen* devoted to chemistry, Webster tried to reconcile the Paracelsian doctrine of principles with van Helmont's teachings along the following lines:

> And though Helmont with the experiments of his Gehennal fire, and some other solid arguments labour the labefaction of this truth, yet doth he not prove that they are not Hypostatical principles, but onely that they are not the ultimate reduction that the possibility of art can produce, which he truly proves to be water; yet are the most compound bodies in the universe to be reduced into them.[85]

WALTER CHARLETON

John Webster's early support of van Helmont's iatrochemistry appears in a work which is in deliberate opposition to university curricula and academic medicine; but the Helmontian translations published in 1650 by Walter Charleton – who was then a candidate for the College of Physicians – show that van Helmont's doctrines were rapidly attracting the interest of part of the medical establishment too. As we have seen with John French, interpretations of van Helmont's ideas were by no means univocal. Sometimes Helmontians focussed on the preparation of new medicines, at other times, the emphasis was on theoretical themes, both in chemistry and

London (Ithaca and London, 1986), p. 122.

[82] Biggs, *Mataeotechnia* (n. 81), p. 123.

[83] Ibid., p. 219.

[84] W. Johnson, *Short Animadversions upon Noah Biggs*, published with L. Fioravanti, *Three Exact Pieces…* (London, 1652). On Johnson, chemist to the College of Physicians, see Cook, *The Decline* (n. 81), pp. 125-6.

[85] J. Webster, *Academiarum Examen* (London, 1654), p. 77.

in medicine. Charleton's early interest in van Helmont focussed on the philosophical motives of the Belgian physician's work.

The intellectual career of Walter Charleton has been described as being influenced by a conversion from his initial adherence to van Helmont's doctrines, to embracing the mechanical philosophy which he espoused in his *Physiologia* of 1654.[86] However, though in the *Physiologia* Charleton did retract his previous adherence to van Helmont's magnetic cure of wounds, there is evidence that no such thing as a total conversion occurred.[87] Indeed, as we shall see, atomism was already adopted in Charleton's own introduction to the Helmontian tracts, while Helmontian notions, and iatrochemical ideas may be found in works published after 1650. Finally, Charleton's version of atomistic philosophy as contained in the *Physiologia*, can hardly be described as purely mechanical.

The first work published by Charleton was the Latin *Spiritus Gorgonicus* of 1650, devoted to the study of stone. Charleton made use of a variety of chemical authors (Severinus, Libavius, Sennert), as well as van Helmont (*De Lithiasi,* 1644). Unlike the latter, Charleton did not reject the Aristotelian elements and qualities. His point of view is that stones are formed by a lapidifying juice – a plastic principle implanted in the Earth – while the four elements operate as subordinate agents.[88] Like Sennert, Charleton advocates a compromise between the Aristotelians and the Paracelsians, and rejects the view that celestial influences are responsible for the generation of *calculi.*[89]

The translation of van Helmont's controversial *De Magnetica Vulnerum Curatione*, which was supported by Charleton's friend Sir Kenelm Digby, is preceded by a lengthy introduction, which contains Charleton's own

[86] See N. Gelbart, 'The Intellectual development of Walter Charleton', *Ambix* 18 (1971), 149-78 and L. Sharp, 'Walter Charleton's Early Life, 1620-1659, and Relationship to Natural Philosophy in Mid-Seventeenth Century England', *Annals of Science* 30 (1973), 311-40. According to Kargon, "Charleton abandoned his earlier adherence to the ideas of van Helmont and became an enthusiastic atomist", Kargon, *Atomism* (n. 3), p. 84. The best intellectual biography of Charleton is S. Fleitmann, *Walter Charleton (1620-1707), "Virtuoso"* (Frankfurt am Main and New York, 1986).

[87] Charleton, *Physiologia Epicuro-Gassendo-Charletoniana* (London, 1654), p. 381.

[88] Charleton, *Spiritus Gorgonicus* (Leiden, 1650), pp. 9-11.On the generation of stones cf. F.D. Adams, *The Birth and Development of the Geological Sciences* (New York, 1954[2]), pp. 77; 136.

[89] Charleton, *Spiritus* (n. 88), pp. 9, 13-14, 46.

explication of the *unguentum armarium*.[90] The cure was based on a balsam (one of whose ingredients was moss grown on the skull of a hanged man) which had to be applied not to the wound, but to the weapon which inflicted it. Like Digby's powder of sympathy, it was meant to heal the wound at a distance. Van Helmont believed that the cure was natural and its operations could be explained as a form of magnetism. The universal spirit operated as a carrier of the curative virtue from the weapon to the wound.[91] Charleton follows van Helmont's view that the cure is natural and is analogous to magnetism, but his own explanation is slightly different from that of the Belgian physician. Charleton explicitly states that he is not confining himself to the translation; he gives his explanation of the puzzling phenomenon of the magnetic cure. He states that action from a distance should not be rejected from natural philosophy: "I am bound to believe that in the infinite magazine of Nature are to be found Agents not obliged to the dull conditions of an immediate Corporeall Contact, but richly endowed with an influentiall or Radiall Activity."[92] Action from a distance is performed by emissions of semi-immaterial atoms, which carry the *unguentum armarium* to the wound.[93] To support his explanation of this strange therapy, Charleton argues that all bodies, but mainly the unctuous ones, incessantly emit a multitude of invisible atoms. Contagion, he continues, is in fact produced by emissions of pestilential atoms into the air.[94] In the 'Translator's

[90] Charleton, *A Ternary of Paradoxes: The Magnetick Cure of Wounds, The Nativity of Tartar in Wine, The Image of God in Man* (London, 1650). I have used the 2nd edn, revised, of 1650; id., *Deliramenta Catarrhi: or the Incongruities, Impossibilities, and Absurdities Couched under the Vulgar Opinion of Defluxions* (London, 1650, published with the 2nd edn of *Ternary of Paradoxes*).

[91] J.B. van Helmont's *De Magnetica Vulnerum Curatione* was first published in 1621 (apparently against the author's will) and then reprinted in *Ortus Medicinae* (1648). The publication of van Helmont's tract led to the trial which ended with his house arrest. See *Propositiones notatu dignae, depromptae ex ejus [Helmontji] Disputationes de Magnetica Vulnerum Curatione Parisiis edita* (Liège, 1624); C. Broeckx, 'Notice sur le Manuscrit Causa J.B. Helmontii', and id., 'Interrogatoires', *Annales de l' Académie Archéologique Belge,* 1852, pp. 277-327 and 306-50; W. Pagel, *van Helmont* (n. 36), pp. 8-13

[92] Charleton, *A Ternary of Paradoxes* (n. 90), sig. d4r.

[93] Ibid., sig. e1r.

[94] Ibid., sig. e3v. For the corpuscular interpretation of contagion, see G. Fracastoro, *De Sympathia et Antipathia rerum, liber unus: De Contagione et contagiosis morbis et eorum curatione, libri III* (Venice, 1546). On the theory of contagion, see V. Nutton, 'The Seeds of Disease: An Explanation of Contagion and Infection from the Greeks to the Renaissance', *Medical History* 27 (1993), 1-34, and

supplement' Charleton highlights the role of magnetism in natural philosophy as, in his view, it provides the key to the understanding of three crucial problems, i.e. the origin of forms, the causes of sympathies and antipathies, and the power of imagination. It should be noticed that the second Helmontian tract translated by Charleton (*Tartari Historia*) criticises a number of views expressed by Paracelsus on tartar. This shows that in his interpretation of van Helmont, he stresses the discontinuity between the thinking of the one and the other – which is in fact also apparent from a number of tracts in the *Ortus Medicinae*. Charleton also explains the generation of tartar along corpuscular lines (an explanation not to be found in van Helmont) by stating that the acid spirit becomes fixed for the action of terrestrial atoms which impede its movements.[95]

Two years after the publication of the Helmontian tracts Charleton sent to press *The Darkness of Atheism*, a work aimed at freeing atomism from its links with atheism. The motion of atoms (as in Gassendi) is not without problems. Though Charleton rejected a number of Epicurean doctrines on atoms (mainly the infinity and eternity of atoms) and stressed the Creationist view of nature and matter, he listed motion among the primary properties of atoms.[96] Contrary to the Aristotelian view that the origin of motion is form and matter is passive, Charleton maintains that matter is active, for God created atoms and endowed them with "a faculty of self-motion", which however God directs.[97] So atoms account for the generation of living bodies, including spontaneous generation. Though Charleton rejects any agent between God and the natural world, he does refer to a plastic spirit or *Archeus* as the formative agent in the generation of animals.[98] Charleton, who seems to be aware of the spreading opposition to official medicine, rejects Paracelsus's macro-microcosm analogy and explicitly distances himself from van Helmont's attacks on academic medicine. He does not deny the existence of the four humours but interprets them in corpuscular

id., 'The Reception of Fracastoro's Theory of Contagion. The Seed That Fell among Thorns?', *Osiris*, 2nd series 6 (1990), 196-234.

[95] Charleton, *A Ternary of paradoxes* (n. 90), sig. f1v.

[96] Charleton, *The Darkness of Atheism Dispelled by the Light of Nature* (London, 1652), p. 46.

[97] Ibid., pp. 47 and 53. For Gassendi's views of matter and motion see above, ch. 2, pp. 64-5. Thomas Hobbes stated that even hard bodies are composed of parts which are in continuous and rapid motion: "Durissima corpora illa sunt in quibus partium motus et velocissimus est, et intra spatia brevissima". *Problemata Physica*, 1662, in T. Hobbes, *Opera Philosophica*, 5 vols. (London, 1839-45), iv, pp. 333-41.

[98] Charleton, *The Darkness* (n. 96), p. 53.

terms.[99] As we shall see, in subsequent works Charleton abandons the theory of humours and adopts the doctrine of spirits.

The *Physiologia* of 1654 stands out as a comprehensive tract of natural philosophy, based entirely on the atomic theory of matter. It is well known that many sections of it are a mere translation of Gassendi's *Animadversiones*.[100] Charleton claims that the four elements are made up of atoms, which have different forms. Though he recognised (as Gassendi did) that it is impossible to establish the exact geometrical figures of atoms and of pores which compose determinate bodies, he did not refrain from establishing the geometrical forms of atoms of fire and cold. Forms determine the propensity or non-propensity to motion: while atoms of fire are small and spherical, atoms of cold ('frigorifick atoms') are tetrahedral or pyramidal. Charleton also suggests that salt is formed of cubical atoms.[101] However, in his explanations of qualities, Charleton had limited recourse to the geometrical forms of atoms. Sometimes he adopted the notion of molecule – which in fact he defined as "seminaries of qualities" – and on most the cases that of contexture of atoms.[102] The action of solvents is in fact explained as follows:

> Every concretion requires to its dissolution some peculiar dissolvent, that holds some respondency or analogy to its contexture.[103]

As in *The Darkness of Atheism*, in the *Physiologia* Charleton adopted the view that atoms were endowed with gravity, that is, an internal principle of motion.[104] Matter is active, so in all bodies, even those having the most compact texture, some atoms are in continuous motion:

> [bodies] are compacted of such principles, as they are indefinently motive, and in perpetual endeavour of emergency or Exilition: so that never desisting from internal evolutions, circumgyrations, and other changes of position; they at length infringe that manner of reciprocal Coaptation, Cohaesion, and Reviction, which determined their solidity, and thereby dissolving the compositum, they wholly emancipate themselves, obey their restless tendency at randome, and disappear.[105]

[99] Ibid., pp. 201, 221.
[100] See Fleitmann, *Charleton* (n. 86), pp. 416-7.
[101] Charleton, *Physiologia* (n. 87), pp. 31-2; 119; 297; 306-7.
[102] Ibid., pp. 109; 119; 266; 271-2; 318-21.
[103] Ibid., p. 266.
[104] Ibid., pp. 112; 121; 124-6.
[105] Ibid., p. 125.

Such a tendency is confirmed by instances taken from chemistry. Charleton mentions spirit of nitre and the actions of various menstrua, including the universal solvent of van Helmont (the *Alkahest*), whose particles are subtle and very active.[106] Indeed, the most active atoms, or rather molecules, form the volatile components of natural bodies, namely their spirits – a notion which plays a central role in subsequent medical works – which carry the gross and less active particles with them.[107] Here Charleton maintains that "the faculties of animals (and the ratiocination of man only excepted) are identical with Spirits, i.e., the most subtile, most free, most moveable or active part of its materials."[108]

Chemistry plays an important role in Charleton's natural philosophy. He maintains that chemical experiments – as he put it "syncritical and diacrytical" – adopted by Sennert provide evidence for the existence of atoms. By means of chemistry "all Bodies are sensibly dissolved into those Moleculae, or first Conventions of Atoms, which carry their specific seminaries."[109] No less important are the 'experimental' arguments of the plants' regeneration from their ashes, like those reported by Libavius, Quercetanus and Gaffarel. Evidently, Charleton believed that in the ashes are atoms endowed with seminal or formative powers.[110]

Like Thomas Willis's and Henry Power's views of spirit, those of Charleton, which are central to his physiology, are largely based on the teachings of Francis Glisson. A look at Glisson's main physiological ideas is therefore necessary.[111]

[106] "But, because our sense, as well as our reason; may have some satisfaction, touching the perpetual commotion of Atoms, even in Compositions; we offer to Exemplifie the same either in the spirit of Halinitre, or that which Chymists usually extract from Crude Mercury, Tin, Sublimate codissolved in a convenient menstruum: For, either of these Liquors being close kept in a luted glass, you may plainly perceive the minute moleculae, or seminary conventions of Atoms, of which it doth consist, to be incessantly moved every way, upward, downward, transverse, oblique, & c. in a kind of fierce aestuation, as if goaded on by their inherent Motor, or internal impulsive Faculty, they attempted speedy emergency at all points, most like a multitude of flyes imprisoned in a glass Vial." Ibid, p. 125. For the *Alkahest*, see ibid., p. 267.

[107] Ibid., p. 270. The section on spirits provides a bridge between the atomistic theory of matter as contained in *Physiologia* and Charleton's medical doctrines.

[108] Ibid., p. 271.

[109] Ibid., p. 109.

[110] Ibid., pp. 109-110. On the regeneration of plants, see F. Secret, 'Palingenesis, Alchemy and Metempsychosis in Renaissance Medicine", *Ambix* 26 (1979), 81-99.

[111] On Francis Glisson (1597-1677), Regius Professor of Medicine at Cambridge

In *De Rachitide* (1650) Glisson made no use of chemistry, but followed the doctrine of temperaments. However, this doctrine was reinterpreted by making it dependent on the activity of vital spirits. Glisson asserted that the slow motion of spirits was the cause of cold temper and, accordingly, of rickets. Blood fermentation is described as the outcome of the spirits' activity, which pass from a state of fixation to one of excitation.[112] In *De Rachitide* Glisson makes only a passing reference to chemical principles, but in his subsequent *Anatomia Hepatis* (1654), where Glisson fully recognises the fundamental role of chemistry for medicine, the five spagyrical principles, that is spirit (or mercury), oil, salt, phlegm (or water) and *caput mortuum* (or earth), are the ultimate components of mixed bodies, including the four humours.[113] Having rejected the traditional definitions of spirit as too vague ("corpus ad volatilitatis gradum rarefactum", or "corpus subtile, activum"), Glisson gives his own definition of this substance, which in fact stresses the importance of two chemical operations, fermentation and distillation, as preliminary to its extraction:

> Spiritus vocabulo intelligendum venit elementum illud, quod, post debitam fermentationem, licet haud antea, sua sponte sursum nititur, & fit volatile. Spiritus hic in distillatione, post praeviam fermentationem primus ascendit. Atque haec est vocabuli spiritus acceptio propriissima, quatenus intelligitur esse misturae pars

(1636-1677), FRS and Fellow and President of the College of Physicians, see O. Temkin, 'The Classical Roots of Glisson's Doctrine of Irritation', *Bulletin of the History of Medicine* 38 (1964), 297-328, W. Pagel, 'Harvey and Glisson on irritability, with a note on van Helmont', *Bulletin of the History of Medicine* 41 (1967), 497-514, John Henry, 'Medicine and Pneumatology: Henry More, Richard Baxter, and Francis Glisson's *Treatise on the Energetic Nature of Substance*', *Medical History* 31 (1987), 15-40. For Henry Power's views of spirits see H. Power, *Experimental Philosophy* (London, 1664), pp. 61-72. Power's iatrochemical theories are in an unpublished essay entitled 'Analogia inter operationes chymicas et naturales', BL, Sloane MS 1393, fols 37^r-50^v, dated 1 May 1657.

[112] F. Glisson, *De Rachitide, sive morbo puerili* (London, 1650), pp. 35-6, 81. The work was the outcome of Glisson's collaboration with G. Bate and A. Regemorter. See E. Clarke, 'Whistler and Glisson on Rickets', *Bulletin of the History of Medicine* 36 (1962), 48-9.

[113] "Quod ad mistorum corporum in elementa ultima divisionem attinet; existimem, sententiam Chymicorum esse verissimam, nempe quinque principia ab iis dicta, puta spiritum, oleum, salem, aquam, sive phlegma, & caput mortuum, sive terram (ut vocant damnatam), esse partes ultimas in quas res ullae vel ingenio, vel industria humana dissolvi queant." F. Glisson, *Anatomia Hepatis* (London, 1654). The quotation is from the 2nd edn, 1659, p. 32. See also ibid., p. 27.

elementaris.[114]

According to Glisson, in natural bodies spirits may be found in three different states, i.e. of fixation, fusion and volatility. Spirits are in the first state when they are closely linked to other principles, in the second when they free themselves from the gross parts of the mixed body, in the third, when they are exalted and can evaporate.[115] As they impart motion to the other principles, spirits are the source of activity in all natural bodies.[116] Glisson ruled out the mechanical explanation of qualities (i.e. the one based on the action on sense organs of atoms having different figures) and maintained that qualities of bodies originate from the combination of spirits with the other chemical principles.[117] Glisson's physiology is based on the vital spirits, which are the most active parts of blood. In blood these vital spirits 'fight' with other chemical principles and from this movement vital heat is generated.[118]

In *The Natural History of Nutrition* Charleton adopted Glisson's tripartition of spirits, as well as the theory that vital heat is produced by the movement of vital spirits.[119] Unlike Glisson, Charleton combined the doctrine of chemical principles with that of atoms. This is apparent in his views on blood. Vital heat is produced by the continuous motion of the most active corpuscles of blood (spirits). Blood is purified by fermentation (i.e. separation of the grosser from the subtle parts) and by the similarity between the size and form of the 'excrementitious' particles with the pores of any part of the body.[120] Charleton's adoption of the shape and size of particles (and pores) does not imply his adherence to mechanical physiology. He explicitly embraced the view that matter is endowed with activity and sensation. As he put it: "all parts of the body have a certain Naturall sense or feeling distinct

[114] Ibid., p. 349. Glisson's adherence to the iatrochemical ideas is attested by some of his manuscripts now in the British Library. Part of Glisson's notes contained in BL MS Sloane 3308 (related to *Anatomia Hepatis*) dealing with the generation of spirits, are published in A. Cunningham, *English Manuscripts of Francis Glisson (1) from Anatomia Hepatis (The Anatomy of Liver) 1654* (Cambridge, 1993).

[115] Glisson, *Anatomia Hepatis* (n. 113), pp. 349-53, 419.

[116] For Glisson, matter is endowed with activity and with perception. See Glisson, *De Natura Substantiae Energetica* (London, 1672).

[117] Glisson, *Anatomia Hepatis* (n. 113), p. 432.

[118] Ibid., p. 366.

[119] Charleton, *The Natural History of Nutrition, of Life, and Voluntary Motion* (London, 1659), pp. 7, 65.

[120] Ibid., pp. 3, 100.

from the animal and wholly independent of the brain."[121]

THOMAS WILLIS

In the *Diatribae duae* (1659) Willis laid the theoretical foundation of his physiology. The first of the two *Diatribae* (*De Fermentatione*), deals extensively with the doctrine of chemical principles – which he interpreted in corpuscular terms. Like Glisson and Charleton, Willis gave a central role to the notion of spirit.[122]

For Willis, mechanical philosophy – which explains natural phenomena by referring them to atoms with different shapes and sizes – deserves praise because it abolished the occult qualities. However he does not adopt it, because – he claims – its principles are presupposed, not proven. In addition, the shape, size and motions of atoms are too remote from observable phenomena to be useful.[123] Willis preferred to adopt the chemical doctrine of principles. This does not imply that he ruled out corpuscular philosophy as such, but he rejected only the mechanical version of atomism.

Willis maintained that, by means of analysis by fire, bodies yield particles of spirit, salt, sulphur, water and earth, which are their components.[124] As we shall see, after the publication of Boyle's *Sceptical Chymist* (1661) Willis changed his view of the five principles. He considered them not as the ultimate constituents of bodies, but just as the product of fire analysis. He endowed the particles of the five principles with chemical qualities and with different degrees of motion. The most active are those of spirits, which play a central part in Willis's natural philosophy and medicine. Sulphur and salt are active principles (though less active than spirits), while water and earth are passive. Willis's division of the five principles into active and passive is by no means original as it emanated from Duchesne and Basso. The particles of the active principles have a strong propensity towards motion. Spirits, we read in *De Fermentatione*, have a natural tendency to motion and they activate the other principles. They are the source of vegetation, of life and the instruments of motion and sensation in

[121] Ibid., pp. 124-125.

[122] Willis, *Diatribae duae medico-philosophicae quarum prior agit de Fermentatione, sive de motu intestino particularum in quovis corpore, altera de Febribus, sive de motu earundem in sanguine animalium. His accessit Dissertatio Epistolica de urinis* (London, 1659). On Willis see H. Isler, *Thomas Willis* (Stuttgart, 1965), Engl. tr. London and New York, 1968 and Frank, *Harvey*, *passim*.

[123] Willis, *De Fermentatione*, in *Diatribae* (n. 122), pp. 3-4.

[124] Ibid., p. 3.

living bodies.[125] Bodies having a small quantity of spirits easily decay as they constitute the *vinculum mixtionis*.[126] Like Glisson and Charleton, Willis explains the activity of spirits according to their three different states. The first is when spirits are dispersed and sluggish and can hardly be extracted; the second when they are more active and can disentangle themselves from grosser particles. The third occurs when spirits can quit the body in which they are contained. Besides motion, the other distinctive property of spirits is their affinity (*cognatio*) with sulphur. The particles of spirit and those of sulphur, according to Willis, form a very stable compound which in turn is the agent of fermentation.[127] Vital spirits originate from a particle of spirit which in the heart is activated by a local ferment.[128] Blood consists of the five principles, which may be found in different proportions and motions. According to Willis, fever is produced by an excessive and irregular fermentation of blood.[129]

CONCLUSION

Before the publication of Robert Boyle's works atomism and chemistry were already linked. Paracelsian ideas were often accepted by atomists and the interpretation of the chemical principles in terms of corpuscles was rather common in England. The atomic theories of matter we find in the first half of the century were not mechanical. Atoms are endowed with chemical qualities, as well as with powers. Descartes's mechanism had little diffusion among philosophers, even less among physicians. The iatrochemists' notion

[125] "Spiritus sunt Substantia maxime subtilis, aetherea, & divinioris aurae particula, quos naturae parens in sublunari hoc mundo, tamquam vitae & animae, motus & sensus cujusque rei instrumenta, condidit; dum sui juris, semper expansi sunt, & avolare nitentes… Ab horum motu corporum animatio, plantarum vegetatio, fructuum, liquorum, & aliorum praeparatorum maturationes procedunt; formam & figuram cuiusvis rei, veluti designatione divina, praefixam, determinant." Ibid., p. 5.

[126] Ibid., p. 5.

[127] "In sinu huius [sulphuris] Spiritus immediate resident, quo velut copula, durioribus coeterorum complexibus, uniuntur.' Ibid., pp. 6-7. "Inter Spiritum & Sulphur est cognatio quaedam & partium similitudo, quae in utroque sunt agiles & dissipatu faciles; quare spiritus e corpore fugatus, particulas sulphureas copiose secum trahit." Ibid, p. 14.

[128] Ibid., p. 25.

[129] Willis, *De Febribus*, in *Diatribae* (n. 122), p. 23. Cf. D.G. Bates, 'Thomas Willis and the fever literature of the seventeenth century', in W.F. Bynum and V. Nutton (eds.), *Theories of Fever from Antiquity to the Enlightenment* (London, 1981), pp. 45-70.

of spirit played an important part in English medicine in the 1640s and 1650s. After the publication of van Helmont's works (1648), iatrochemistry was widespread in England, while the impact of iatromechanism was negligible. As we shall see, Helmontianism played a central part in Boyle's early chemical studies. It also provided arguments to Boyle's criticism of the Paracelsian principles.

CHAPTER 4

ROBERT BOYLE'S CORPUSCULAR PHILOSOPHY

INTRODUCTION

One of Boyle's main scientific pursuits was to explain chemical phenomena in corpuscular terms and to establish chemical foundations for corpuscular philosophy. This project was not entirely original. As we have seen in the previous chapters, a number of chemists and natural philosophers at the beginning of the seventeenth century combined chemistry and corpuscular theories. However, unlike his predecessors, Boyle's combination of chemistry and corpuscular philosophy was based on an articulate theory of matter and was supported by a substantial amount of experimental evidence.

In order to understand Boyle's fusion of chemistry and corpuscular philosophy, it is useful to elucidate his theory of matter, which is commonly described as purely mechanical. As we shall see, though Boyle maintained that mechanical theories are the most intelligible and the simplest which a naturalist can employ, he was reluctant to adopt a straightforward mechanical theory of matter, based on the shape, size and motion of particles of inert matter. This becomes apparent when he deals with chemical and physiological phenomena. He often employed theories based on the chemical, not the mechanical properties of corpuscles.

According to the view held by most historians of science, Boyle was the one who opposed Paracelsianism and reconstructed chemistry on rational (that is mechanical), grounds. This view was expressed in two influential studies, both published in the 1950s: Thomas Kuhn's article of 1952 on Boyle and structural chemistry and Marie Boas's monograph on Boyle published in 1958. According to Boas, "[Boyle's] chemistry was sufficiently theoretical, but also [...] rational and mechanistic to be regarded as worthy of inclusion in the new experimental natural philosophy."[1] Thomas Kuhn stated that "Boyle's faith in the corpuscular principles of the mechanical philosophy is the cause of his emphasis in chemistry upon structure, configuration and motion, as well as a cause of his rejection of explanations

[1] M. Boas, *Robert Boyle and Seventeenth-Century Chemistry* (Cambridge, 1958) (hereafter Boas, *Robert Boyle*), p. 67.

in terms of inherent characteristics of the ultimate corpuscles." His chemistry, said Kuhn, was based on mechanical philosophy, and for this reason it was "incompatible with the belief in the existence of enduring elements."[2]

This interpretation of Boyle's theory of matter as strictly mechanical had not been upheld by Boyle's contemporaries. It emerged with a clear philosophical meaning at the beginning of the eighteenth century. It was formulated by Leibniz in the context of his own rejection of the Newtonian concept of force. Leibniz fostered this image of Boyle as a strict mechanical philosopher, and drew a sharp contrast between the latter's and Newton's (and the Newtonians') theory of matter. In September 1710, *Acta Eruditorum* published a review of John Freind's *Praelectiones Chymicae.*[3] The author was Christian Wolff, but Leibniz contributed the sentences concerned with Boyle's philosophy. The target of the review was John Keill, but it is evident that the attack was ultimately directed against Newton. Leibniz claimed that Keill, by reintroducing into chemistry occult qualities (the forces operating among the particles of bodies), was destroying the sound mechanical philosophy which Boyle had established and applied to chemistry.[4] This interpretation of Boyle as the mechanical

[2] T.S. Kuhn, 'Robert Boyle and Structural Chemistry', *Isis* 43 (1952), 12-36.

[3] *Acta Eruditorum,* September 1710, 412-16. It is apparent that Freind was aware of the role played by Leibniz in the attack launched on his theories. In the 'Appendix Containing the Account given of these Lectures in the Lipsick Acts, together with some Remarks thereon', published in the English translation of his *Praelectiones,* Freind's reply was mainly directed against Mr L[eibniz], who he claimed inspired the views contained in the *Acta Eruditorum.* See J. Freind, *Chymical Lectures* (London, 1712), pp. 189-200. On Freind and the Newtonian chemists see A. Thackray, *Atoms and Powers* (Cambridge, Mass., 1970). The work in question is J. Keill, 'Epistola ad Cl[arissimum] virum', *Philosophical Transactions,* 315 (1708), 97-110. It was reviewed in the *Acta Eruditorum* of 1709. For J. Keill, see A. Guerrini, 'The Tory Newtonians: Gregory, Pitcairne and their Circle', *Journal of British Studies* 25 (1986), 288-311.

[4] "Verum enim vero Dn. Keilius cum sequacibus redit reapse ad qualitates occultas, quales apud Scholae Philosophos sympathia & antipathia fuere, dum vim quandam attractricem statuit, quae si (ut ipse vult) primitiva est, omnique materiae erga omnem materiam essentialiter competit, utique per rationes mechanicas explicare nequit, atque adeo vel erit aliquid absurdum, vel in miraculum seu voluntatem Dei extraordinariam resolvetur, ad quam tamen in Physicis sine necessitate confugiendum non esse, convenit inter intelligentes. Quodsi aliter procedimus & fictionibus indulgemus, reditur ad Philosophiam quandam

philosopher *par excellence* was also upheld by Leibniz in his correspondence with Clarke. As we read in Leibniz's fifth paper: "Mr Boyle made it his chief business to inculcate, that every thing was done mechanically in natural philosophy."[5]

In the eighteenth century Boyle was regarded as a physicist rather than as a chemist. In the article 'Chymie' for the *Encyclopédie* written by the Stahlian chemist Gabriel François Venel, Boyle is interpreted as the scientist who promoted mechanical philosophy rather than chemistry. Venel established clear boundaries between chemistry and physics, and complained that Boyle "est trop exactement physicien corpusculaire-méchanicien, ou physicien proprement dit", and suggested placing him among the physicists, rather than among the chemists.[6] Nineteenth-century histories of chemistry recognised Boyle's specific contributions to chemistry, whereas in the twentieth century the view of Boyle as the champion of mechanical philosophy prevailed, though his experimental contributions to chemistry were not neglected.[7]

phantasticam Scholae vel etiam Enthusiasticam, qualis *Fluddi* fuit. Ita uno ictu subvertentur, quae in Anglia ipsa *Robertus Boylius* & alii Viri docti de rebus naturalibus mechanice, id est, rationabiliter explicandis magno studio stabiliverunt, quae Boylius etiam diserte ad Chymica applicuit." (*Acta Eruditorum,* September 1710, 412-3.) This passage is Leibniz's. It is noticeable that in 1692 Leibniz shared Huygens's evaluation of Boyle's scientific work as mainly experimental, having no significant theoretical achievement: "Mr. Boyle est mort, comme vous scaurez desia sans doute. Il paroit assez etrange qu'il n'ait rien basti sur tant d'experiences dont ses livres sont pleins..." Christian Huygens to G. W. Leibniz, 4 February 1692, *Oeuvres Complètes de Christian Huygens,* edited by the Societé Hollandaise des Sciences, 22 vols. (The Hague, 1888-1950), x, p. 239. Leibniz replied on 19 February of the same year: "[...] mais ce que vous dites de feu Mr. Boyle, est ancor veritable à son egard, qu'il n'estoit pas capable d'une assez grande application pour pousser les consequences autant qu'il faut." Ibid., vol. 10, p. 263. It is apparent that Leibniz's subsequent emphasis upon Boyle's mechanical philosophy was mainly aimed at reinforcing his own anti-Newtonian arguments.

[5] *The Leibniz-Clarke Correspondence,* edited by H.G. Alexander (Manchester, 1956), p. 92.

[6] *Encyclopédie, ou dictionnaire raisonné des sciences, des arts et des métiers, par une société des gens de lettres,* 17 vols. (Paris, 1751-1765), iii, p. 435. On Venel see E.M. Melhado, 'Chemistry, Physics, and the Chemical Revolution', *Isis* 76 (1985), 195-211 (esp. pp. 196- 9). A view similar to Venel's is to be found in the *Histoire de l'Académie des Sciences*, 11 vols. (Paris, 1729-34), i, p. 79.

[7] J.C.F. Hoefer, *Histoire de la Chimie depuis les temps les plus reculés jusqu'à notre époque,* 2 vols. (Paris, 1842-43), ii, pp. 146- 76 and H. Kopp, *Geschichte der*

In the present chapter I wish to demonstrate that: 1. Boyle's theory of matter was corpuscular, but not strictly mechanical, i.e. based on the shape, size and motion of particles of inert matter; 2. Boyle did not consider chemistry as a branch of physics, since he did not reduce all chemical phenomena to the geometrico-mechanical affections of the particles of inert matter; 3. Boyle questioned the spagyrical principles, but did not rule out the existence of simple and homogeneous chemical substances. He denied this title to those bodies (three or five) which chemists commonly believed to be the ingredients of all mixed bodies. He also questioned the current classification of chemical substances and proposed different criteria to classify them.

1. Although Boyle often repeated that the mechanical properties of corpuscles were to be regarded as the most general notions of natural philosophy, a closer analysis of his natural philosophy reveals a number of agents not operating according to the principles of mechanical philosophy. These agents are seminal principles, spirits and ferments – which Boyle conceived as corpuscles endowed with the power of fashioning other parts of matter. The notion of *semina* is of special importance to the understanding of Boyle's theory of matter, as it was adopted to link his corpuscular philosophy to the teleological view of nature. The notions of spirit and ferment, which he used in *The Usefulnesse of Experimental Philosophy* (as well as in the works of the maturity, as for instance the

Chemie, 4 vols. (Brunswick, 1843-1847), i, pp.168-9. Lasswitz, *Geschichte*, ii, pp. 261-300; F.A. Lange, *Geschichte des Materialismus und Kritik seiner Bedeutung in der Gegenwart,* 2 vols. (Leipzig, 1908), ii, pp. 255-9. For the interpretation of Boyle's chemistry as purely mechanical, see E. Bloch, 'Die antike Atomistik in der neuren Geschichte der Chemie', *Isis* 1 (1913), 377-415 (esp. pp. 389-404); Metzger, *Doctrines Chimiques*, p. 234; R.S. Westfall, *The Costruction of Modern Science. Mechanism and Mechanics* (Cambridge, 1977), p. 76; R.H. Kargon, *Atomism in England from Hariot to Newton* (Oxford, 1966), pp. 93-105, esp. p. 100, denied the originality of Boyle's corpuscular philosophy (which he saw as a synthesis of Gassendi's and Descartes's views) and restricted Boyle's contribution to the theory of matter to the experimental support he sought for the mechanical philosophy. Hooykaas, 'Het Begrip', pp. 195-201 and E.J. Dijksterhuis, *The Mechanisation of the World Picture* (Oxford, 1961), pp. 435-7 gave a more perceptive interpretation of Boyle's theory of matter, stressing the importance of Boyle's notions of compound corpuscles and of texture. A good account of Boyle's notion of texture is contained in P. Alexander, *Ideas, Qualities and Corpuscles. Locke and Boyle on the External World* (Cambridge, 1985), pp. 66-7 and 85-6. Partington, ii, pp. 486-549, contains the most detailed account of Boyle's chemical experiments.

History of Blood), testify to Boyle's debts to Helmontian iatrochemistry. The notion of ferment occurs also in Boyle's alchemy. He described the elixir as a ferment, a substance which has the power of transmuting a huge quantity of matter, many times its weight.[8]

2. In order to assess the status of chemistry in Boyle's work and its relationship to mechanical philosophy, we have to consider the role of mechanical principles in Boyle's explanation of natural phenomena. It is well known that Boyle considered that explanations based on the shape, size and motion of corpuscles were the primary, simplest and most comprehensive a naturalist could adopt.[9] However, he did not deduce all phenomena from the primary affections of corpuscles (shape, size and motion). As he put it, "there are so many subordinate causes between particular effects and the most general causes of things, that there is left a large field, wherein to exercise mens industry and reason."[10] This statement is not isolated. There are analogous statements both in Boyle's published works and in his manuscripts. His adoption of what he called "intermediate theories" marked a significant departure from the strict mechanical philosophy. Unlike Descartes, Boyle did not consider that explanations based on the shape, size and motion of the primary corpuscles were the only valid ones. Indeed, he seldom referred to the "catholick affections of matter" when he dealt with chemical phenomena. Boyle had recourse to compound corpuscles, namely corpuscles endowed with chemical, not only mechanical, properties. Boyle's interpretation of chemical phenomena was ultimately based on his classification of corpuscles, which will be discussed below.

3. Boyle's articulated refutation of the Aristotelian elements and of the spagyrical principles contained in *The Sceptical Chymist* has been interpreted as a sign of Boyle's reluctance to assess any positive chemical theory at all.[11] On closer analysis, it is apparent that Boyle did not preclude

[8] As Principe pointed out, "Boyle adopted parts of alchemical theory for use as explanatory principles – sometimes in conflict with his now more-celebrated mechanical principles" 'Boyle's Alchemical Pursuits', in *Robert Boyle Reconsidered*, p. 97.

[9] Boyle, *Of the Excellency and Grounds of the Mechanical Hypothesis*, published as Appendix to *The Excellency of Theology* (London, 1674), *Works*, iv, pp. 70-1.

[10] Boyle, *The Usefulnesse*, *Works*, ii, p. 45.

[11] Cf. R-M. Sargent, *The Diffident naturalist. Robert Boyle and the Philosophy of Experiment* (Chicago, 1995), p. 53.

the possibility of discovering simple and homogenous chemical substances. What Boyle attacked was the Paracelsians' view that a small number (three or five) of simple substances are the ingredients of *all* bodies and can be extracted from *all* mixed bodies by means of fire analysis. The main points of disagreement with the Paracelsians' theory of principles were the following: first, he rejected the view that all natural bodies contain the same substances; second, he questioned the validity of fire analysis, which in his view does produce new substances, but does not yield the original ingredients of compound bodies. However, Boyle believed that a powerful solvent, like van Helmont's *Alkahest*, could disclose the constituents of mixed bodies. Therefore, Boyle's criticism of the Paracelsian doctrine of the *tria prima* (and of van Helmont's water) does not mean that he regarded all chemical substances as compound bodies. He did not deny that simple and homogeneous substances could be discovered, though he never tried to assess their number. It is however apparent that such a quest was not central to his chemical research. More relevant to Boyle's chemical investigations was the reassessment of the chemists' classification of chemical substances. Boyle did not rule out the possibility of a classification of substances based on their chemical properties. His aim was to increase the number of what he called "chemical families", as, in his view, chemists had based their classifications on a few similarities which various substances showed, ignoring the differences which could be made manifest only if one forced them to appear by devising appropriate experiments.

As regards the chemical properties, it is necessary to clarify Boyle's own position, which historians have often oversimplified. Boyle's criticisms of the chemists' doctrine of qualities did not mean that he saw all chemical properties as immediately reducible to the mechanical attributes of the particles of matter. Indeed, Boyle's aim was to reject the chemists' notion that sensible qualities were reducible to a given principle. On the other hand, he refrained from establishing a direct relationship between a given quality and a set of mechanical properties of the simplest corpuscles. As we shall see, Boyle maintained that qualities had relative character. They stemmed from the various interactions of different corpuscles which themselves might not carry the quality in question.

THE BACKGROUND TO *THE SCEPTICAL CHYMIST*

In the late 1640s and early 1650s Boyle developed his chemical studies in conjunction with the Hartlib Circle. His correspondence and personal contacts with Clodius, Starkey and Worsley show mutual interest in a number of chemical and alchemical topics. There is also evidence that Sir Kenelm Digby, himself linked with the Hartlib Circle, exerted an important influence on Boyle's early chemistry. In these years Boyle became familiar with a mass of chemical and alchemical texts, and became acquainted with the works of contemporary chemists such as Rudolph Glauber and several French chemists.[12]

Signs of Paracelsus's (and of the Paracelsians') influence can be traced in Boyle's early chemistry. At the beginning of the 1650s, Jean Baptiste van Helmont's iatrochemistry, which, as we have seen, was widely known in the Interregnum, became a major source of Boyle's chemical investigations. *The Usefulnesse* and the so-called first draft of *The Sceptical Chymist* are largely based on Helmontian theories and experiments. Though in *The Sceptical Chymist* Boyle criticises van Helmont's theory of water and *semina* as the principles of natural bodies, crucial Helmontian views are still adopted there and in subsequent works.

Boyle's diary of January 1649 (old style) contains notes on ferments and on chemical spirits, two topics which played a prominent part in Paracelsian and Helmontian iatrochemistry.[13] Part I of *The Usefulnesse*, largely written

[12] It is difficult to assess when Boyle's commitment to chemistry started. M. Hunter maintained that before 1648 "Boyle was a moralist, not a scientist". (M. Hunter, 'How Boyle became a Scientist', *History of Science* 33 (1995), 59-103. The quotation is from p. 63. See also J. Harwood (ed.), *The Early Essays and Ethics of Robert Boyle* (Carbondale and Edwardsville, 1991); L. Principe, 'Style and thought of the early Boyle: Discovery of the 1648 manuscript of Seraphic Love', *Isis* 85 (1994), 247-60 and id., 'Virtuous Romance and Romantic Virtuoso: The Shaping of Robert Boyle's Literary Style', *Journal of the History of Ideas* 56 (1995), 377-97. For a different view of Boyle's early career, see M. Oster, 'Biography, culture and science: The formative years of Robert Boyle', *History of Science* 22 (1989), 151-79. A letter from Robert Boyle to his sister Katharine Ranelagh dated 6 March 1646/7 unambiguously testifies to Boyle's early chemical studies, see *Works*, i, pp. xxxvi-xxxvii. On Frederick Clodius see Webster, *Instauration*, pp. 302-4. On Benjamin Worsley, see DNB.

[13] Boyle's correspondence with Oldenburg of 1659 contains evidence of his interest in the views of French chemists (i.e. Rochas and Nuysement) on the spirit of the world and the way to fix it. See Oldenburg, *Correspondence*, i, pp. 214, 233-4 and 245-6. Undated manuscript notes show that Boyle planned to write a natural history of spirits (Royal Society Commonplace book, MS 186, fol. 19^r). See A. Clericuzio, 'The Internal Laboratory. The chemical reinterpretation of medical spirits in England (1650-1680)', in P. Rattansi and A. Clericuzio (eds.), *Alchemy*

between 1648 and 1650, shows evidence of Boyle's commitment to iatrochemistry and to Paracelsianism. In the first essay Boyle extols Paracelsus and explicitly criticises Galenic physicians. Unlike most of the latter, Paracelsus – despite his "many extravagances" – improved our knowledge of nature and produced useful remedies.[14] Boyle commends both Paracelsus's inquisitive attitude and the Paracelsian view that direct investigations of nature disclose God's power in the world. Like other members of the Hartlib Circle, Boyle was convinced that Paracelsianism contributed both to the advancement of learning and to the promotion of the Christian religion.[15] The first essay of *The Usefulnesse* also contains a short account of Boyle's presumably extensive investigations of the nature and preparation of antimony – a substance which Paracelsians had introduced into medicine.[16] When discussing the relationships of chemistry to medicine, Boyle evidently adopted the Paracelsians' view, namely that chemistry provides the foundations to medicine. Chemistry, as he put it,

> assists us, by the resolution of bodies, to extricate their more active parts, and partly by such resolutions and partly by associating bodies together, to alter the former texture of nature's productions, or present us with new concretes of new textures; by this very means, if men want not curiosity, and industry to vary and prosecute experiments, there must necessarily arise such a store of new and active medicines, that, in all probability, many of them will be found endowed with such virtue, as have not been, at least in that degree, met with in the usual medicines, whether simple or compound.[17]

and Chemistry in the 16th and 17th Centuries (Dordrecht, 1994), pp. 51-83 (esp. p. 56).

[14] *The Usefulnesse*, *Works*, ii, pp. 13-4. On Boyle's attitude towards Galenic medicine, see M. Hunter, 'Boyle versus the Galenists: a Suppressed Critique of Seventeenth-Century Medical Practice and its Significance', *Medical History* 41 (1997), 322-61.

[15] *The Usefulnesse*, *Works*, ii, p. 57. Cf. Webster, *Instauration*, *passim*.

[16] *The Usefulnesse*, *Works*, ii, p. 11. Boyle's list of tracts on antimony he had used contains the relevant ones on the subject: Alexander von Suchten, *Liber unus de secretiis antimonii* (1570), Angelo Sala, *Anatomia Antimonii* (1617), Hamerus Poppius, *Basilica Antimonii* (1618); Basilius Valentinus, *Currus triumphalis Antimonii* (1646).

[17] *The Usefulnesse*, *Works*, ii, p. 149.

Like the Paracelsians and the Helmontians, Boyle maintained that a number of chemical reactions take place in the human body. This is the case with digestion, which he interpreted in chemical, not in Galenic terms:

> and it seems a mistake to imagine (how many soever do so) that heat must needs be the efficient of all the changes the matter of our aliments may happen to undergoe in a humane body: where there are strainers, and solvents, and new mixtions, and perhaps ferments, and divers other powerfull agents, which by successively working upon the assumed matter, may so fashion and qualifie it, as in some cases, to bring the more disposed part of it to be not unlike salts or other mineral substances.[18]

Though Boyle does not share the iatrochemists' view that fermentation is the only cause of heat and effervescence in human bodies, he highlighted the importance of fermentation to both physiology and to pathology. As we gather from *The Usefulnesse*, Boyle had planned to write a tract specifically devoted to fermentation.[19]

At the beginning of the 1650s Boyle became more critical towards Paracelsianism. He criticised crucial aspects of Paracelsian iatrochemistry, namely, the doctrine of principles and the related view of fire analysis, the Paracelsians' overall rejection of anatomy and the so-called micro-macrocosm analogy.[20] Much of Boyle's criticisms of Paracelsianism were rooted in van Helmont's works. We can safely affirm that in the 1650s Boyle adhered to central aspects of van Helmont's iatrochemistry, and at the end of the decade, as attested by *The Sceptical Chymist* (1661), he rejected van Helmont's theory of principles. Nonetheless, Helmontian ideas and experiments can still be found in the works of Boyle's maturity.

Part two of *The Usefulnesse*, which is entirely devoted to medicine, provides the strongest evidence of Boyle's favourable reception of Helmontian iatrochemistry. As witnessed by his words of advice for the

[18] Ibid., p. 82. See also id., pp. 81-2: "though some Paracelsians do take too much liberty, when they crudely tell us, that there are arsenical, vitriolate, aluminous, and other mineral substances, generated in human bodies; yet, if they had more warily proposed their doctrine, it would not perhaps appear so absurd, as they are wont to think it, who considering only the nature of the aliments men usually feed upon, cannot conceive, that such being but either animals or vegetables, can by so gentle a heat as that of man's body [...] be exalted to an energy like that of such bodies, as are composed of active mineral substances."

[19] Ibid., p. 83. It would seem that this tract was never written.

[20] Ibid., p. 54.

advancement of medicine, Boyle fully embraced van Helmont's view that chemistry was the foundation of medicine. The list of subjects Boyle recommends to the physicians is largely dependent on van Helmont. There is however one exception: that Boyle does not rule out entirely the theory of humours – at least at this stage of his career. His iatrochemical agenda clearly shows the Helmontian leanings of Boyle. It includes the study of the saline and sulphureous parts of the various liquors contained in the body; of the acid juices of the stomach; of the spirit of blood; and of the acid-alkali reactions in the body.[21]

As I have investigated Boyle's attitude towards van Helmont's iatrochemistry elsewhere, I am here focusing on issues related to chemical theories and in particular to the principles.[22]

An important part of Boyle's early chemical investigations was the search of the *Alkahest*, van Helmont's universal solvent. The quest for the *Alkahest* was also part of Starkey's and Clodius's chemical research. As Boyle became increasingly dissatisfied with traditional chemical analysis, he regarded the *Alkahest* as an instrument which could better disclose the composition of bodies than the analysis by fire. The quest for the *Alkahest* never disappeared from Boyle's agenda. In *The Sceptical Chymist* he positively expressed himself on the existence of the *Alkahest*, or rather of a similar solvent:

> And I have heard from very credible eye-witnesses some things, and seen some others myself, which argue so strongly, that a circulated salt, or a menstruum (such as it may be) may by being abstracted from compound bodies, whether mineral, animal, or vegetable, leave them more unlocked than a wary naturalist would easily believe.[23]

Boyle's interest in the production of the universal menstruum has at least two motivations. First, he hoped that such solvents could open the textures of bodies, especially of metals – a feature of van Helmont's universal

[21] Ibid., pp. 79-80.

[22] See A. Clericuzio, 'Van Helmont's iatrochemistry and the Hartlib Circle', in S. Mandelbrote (ed.), *The Hartlib Papers: A Universal Correspondency*, forthcoming.

[23] *The Sceptical Chymist* (London, 1661), *Works*, i, p. 486. *Sal circulatum* was Paracelsus's universal solvent, see L. Reti, 'Van Helmont, Boyle and the Alkahest', in L. Reti and W.C. Gibson, *Some Aspects of Seventeenth-Century Medicine and Science* (Los Angeles, 1969), pp. 3-19; Newman, *Gehennical Fire*, pp. 146-8 and 181-8 and B. Joly, 'L'Alkahest, dissolvant universel ou quand la théorie rend pensable une pratique impossible', *Revue d'histoire des sciences* 49, 2-3 (1996), 305-44.

menstruum. Second, he sought a solvent which could be recovered after its operations – which again was one of the properties van Helmont attributed to the *Alkahest*. As Boyle put it, "the solvent should not be so spoiled by a single operation, made with them, as our vulgar saline spirits are wont to be..., but being drawn off from the dissolved body, or the extraction, will again serve, more than once, for the like operation upon fresh materials."[24]

Though Boyle hoped that the *Alkahest* could help disclose the ultimate ingredients of natural bodies, he did not share the view that the outcome of the *Alkahest*'s operation would be the reduction of bodies to one simple and homogeneous substance (which for van Helmont was water). It is apparent that Boyle did not put forward hypotheses about the kind of substance the *Alkahest* could extract.[25]

The most informative piece of evidence of Boyle's chemical theories in the mid-1650s is a manuscript entitled 'Reflexions on the Experiments vulgarly alledged to evince the 4 Peripatetique Elements, or ye 3 Chymicall Principles of Mixt Bodies', published by Marie Boas as an early draft of *The Sceptical Chymist*.[26] As the title indicates, Boyle's unpublished tract deals with a topic which was becoming central to his chemical work – the principles of compound bodies. It contains a detailed rejection of both the Aristotelian and the Paracelsian theory of matter. Boyle's arguments against the chemical principles are almost entirely borrowed from van Helmont. Moreover, the 'Reflexions' show Boyle's favourable opinion of some relevant Helmontian views. Following van Helmont, Boyle claims that the

[24] *The Usefulnesse*, *Works*, ii, pp. 143-4.

[25] The recipe of the *Alkahest* was, of course, a matter of contention. We do not know which procedures Boyle had tried to obtain the universal solvent. Starkey's views, as Newman has documented in detail, changed over time. He initially thought that by means of van Helmont's *offa* (ammonium chloride) and vinegar (acetic acid) one could produce the *Alkahest* – a suggestion he communicated to Boyle in 1651. Then he rejected vinegar as an ingredient. Later, in *Liquor Alkahest* (published posthumously in 1675) Starkey became rather pessimistic about the possibility of obtaining the *Alkahest*. See Newman, *Gehennical Fire*, pp. 179-88. From a letter of Clodius of 4 July 1654 to an anonymous correspondent we gather that the former rejected the view that the main ingredient of the universal solvent was mercury. HP 16/1/7A-B.

[26] M. Boas Hall, 'An Early Version of Boyle's *Sceptical Chymist*', *Isis* 45 (1954), 153-68 (hereafter 'Reflexions'). M. Boas Hall, *Robert Boyle*, p. 39, maintained that this manuscript was written not later than 1655. The manuscript is also discussed by C. Webster, 'Water as the Ultimate Principle of Nature: The Background to Boyle's *Sceptical Chymist*', *Ambix* 13 (1966), 96-107.

substances which fire analysis yields were not pre-existing in the mixed body, but were produced by fire.[27] Indeed, Boyle's own criticism of the Paracelsians' principles was far more radical than van Helmont's. Boyle questioned fire analysis as the way to obtain simple and homogeneous substances from mixed bodies. His arguments against fire analysis are based on the particulate theory of matter, and notably on the notion of texture:

> For, I consider, that the genuine property of heat is, to dissociate the parts of bodies, and to subdivide them into small particles, without regard to their being homogeneous or heterogeneous [...]. And even, when the fire seems most to congregare homogenea et disgregare heterogenea it produces that effect but by accident, for the fire does but dissolve the Cement or Contexture, that kept the heterogeneous parts of bodies together under one common forme, upon which dissolution the component particles of the mixt being freed and set at liberty, doe naturally, and not by any operation of the fire, associate themselves each with its like, or rather, take those places, which their several degrees of gravity and levity, fixtnes or volatility (either natural or adventitious from the atoms of the fire) assigne them.[28]

As this passage shows, the kind of corpuscles Boyle refers to are endowed with gravity, volatility and fixedness. Boyle's objections to the chemical doctrine of principles are twofold: 1. There are substances which fire cannot decompose into the three principles; 2. the substances fire extracts are not homogeneous. Both objections are fully articulated in *The Sceptical Chymist*.

As far as the first objection is concerned, it is to be observed that, when arguing that the *tria prima* cannot be extracted from gold, Boyle does not deny that someone has obtained sulphur and mercury from gold. It would seem that Boyle's positive view of the two principles of gold was also supported by what Sir Kenelm Digby reported to him.[29] Another passage of

[27] 'Reflexions', p. 159. Cf. van Helmont, 'Complexionum atque mixtionum elementalium figmentum', § 10, *Ortus Medicinae*, p. 105.

[28] 'Reflexions', p. 160. This argument is restated in *The Sceptical Chymist.*

[29] "But, in the next place, I could never see gold by fire divided into so many, as 3 elementary heterogeneities, salt, sulphur and mercury. Tis not, that I dare peremptorily deny, that out of Gold a kind of sulfur may be extracted, leaving the remaining body deprived of its wonted colour; nor, that there may not be drawn out of Gold a real mercury (Sir K. Digby having told me, he hath done the latter) but for a salt of Gold, I could never be satisfied, it was ever separated in rerum natura by any credible Eyewitness: and if it should succeed by those costly materials, they

this manuscript testifies that Boyle considered mercury as a homogeneous substance. If mercury is exposed to the action of fire, it is divided into parts which are of the same nature, since mercury's parts are not dissimilar. Boyle set out to investigate whether water was a simple and homogenous substance – as van Helmont maintained.[30] The Helmontian theory of water and seminal principles is thoroughly discussed in the 'Reflexions'. Unlike in *The Sceptical Chymist*, in the 'Reflexions' Boyle is still rather favourable to the Helmontian theory of two principles. For Boyle, the water theory was confirmed by the chemical analysis of plants and animals, as well as by the 'water culture' experiment which he had successfully repeated.[31] The experimental support to the Helmontian theory of principles did not prevent Boyle from voicing some reservations on the water principle, which in fact was to be developed in *The Sceptical Chymist*. The first is that there is no evidence to prove that minerals and metals are produced by water. The second (and stronger) objection is that, provided that water can be obtained by the *Alkahest*, van Helmont cannot prove that water is a simple and homogeneous substance. According to van Helmont, water is an element because it is insipid. Such an argument, according to Boyle, cannot support the Helmontian doctrine.[32] As far as the other Helmontian principle (*semina*) is concerned, Boyle's 'Reflections' contain no substantial objection:

> I must admire the strange power of the formative power of the seeds of things, which doe not only fashion the obsequious matter according to the exigency of their own natures, and the parts they are to act; but doe also dispose and change the matter, they subdue, as to give it a consistency, which it seemed incapable of admitting.[33]

Boyle adopted the Helmontian seminal principles theory and reinterpreted it in corpuscular terms:

> we may observe in Eggs, where the seminal particles, tho at first scarce discernible to the Eye, lodged as it were between the coates of yolk and white, doe not only prepare the matter into that variety of

talke of, the extraction of this golden salt being to be effected by corrosif menstruums, or the intervention of other saline bodies, it will remaine doubtful, whether the emergent salt be that of the gold itself, or of the saline bodies, or spirits, emploied to prepare it.", 'Reflexions', pp. 161-2.

[30] 'Reflexions', p. 159.

[31] 'Reflexions', pp. 165-6.

[32] 'Reflexions', p. 165

[33] 'Reflexions', p. 167.

> contextures and Consistencies, that is requisite to the production of veine, sinew, artery, tendon, flesh, membrane… ; but dos also out of the same matter produce the bones, so much harder then that soft and liquid substance, whence they are made.[34]

Before considering the way Boyle applied corpuscular philosophy to the study of chemistry, and Boyle's arguments against the chemical theories of principles, I now turn to investigate the origin and development of his corpuscular theory of matter.

THE ORIGIN OF BOYLE'S CORPUSCULAR PHILOSOPHY

The investigation of the origin of Boyle's corpuscular philosophy is not aimed just towards detecting Boyle's sources and towards determining the different phases of development of his theory of matter. It may also elucidate some central aspects of Boyle's theory of matter, such as the classification of corpuscles and the attribution of chemical properties to corpuscles.

The first piece of evidence of Boyle's early views of matter is a manuscript on atomism ('Of ye Atomicall Philosophy'), which he left unfinished. I have elsewhere tentatively dated (1651-1653) this manuscript and have tried to explain why Boyle wanted it to be destroyed.[35] This manuscript was also examined by W. Newman, who stressed the importance of Sennert as the main source for Boyle's early atomism.[36]

Though at the outset of the manuscript Boyle praises the philosophy of Leucippus, Democritus and Epicurus, his view of atoms diverges in many aspects from classical atomism. Evidently Boyle's purpose here is to support atomism with experimental arguments. For him, the best ones were those supplied by chemistry. In fact, Boyle's atoms are rather different from Democritus's and Epicurus's, resembling the *minima naturalia* chemists

[34] 'Reflexions', p. 167. This passage seems to imply Boyle's adherence to the theory of epigenesis. In manuscript notes on spontaneous generation, possibly written at the end of the 1650s, which will be considered in the next paragraph, Boyle follows the theory of preformation.

[35] BP, xxvi, fols 162-175, to be published in M. Hunter and E.B. Davis, *The Works of Robert Boyle*, 14 vols. (London 1999-2000), vol. xiii. Cf. A. Clericuzio, 'A Redefinition of Boyle's chemistry and corpuscular philosophy', *Annals of Science* 47 (1990), 568-9 (n. 34 for the date of composition).

[36] W.R. Newman, 'The Alchemical Sources of Robert Boyle's Corpuscular Philosophy', *Annals of Science* 53 (1996), 567-85.

adopted in the early part of the seventeenth century. The importance Boyle attached to chemical experiments in order to prove the existence of atoms clearly shows that his chemical theories and corpuscular philosophy were in fact already strictly linked at the outset of his career. What we find in subsequent works is a development and an articulation of his theory of matter. In his manuscript notes on atomism Boyle adopts a view of atoms which is almost the same as the *minima naturalia*, that is, a qualitative version of atomism. There is however a difference between the scholastic (and between Sennert's) version of *minima* and the one we find in Boyle's manuscript. The difference is that Boyle rules out the Aristotelian notion of form, which, as we have seen, plays a significant part in Sennert.[37]

In this manuscript the term *minima naturalia* denotes atoms endowed with qualities, while in the published works the same term means simple corpuscles having purely mechanical properties. However, what is important to our purpose is that in this manuscript, as in the subsequent works, Boyle explains chemical reactions by means of corpuscles endowed with chemical properties. As we shall see, in published works Boyle resorts not to the ultimate blocks of matter, but to corpuscles of a higher order of composition. In Boyle's definition of atoms, in the manuscript under scrutiny, we do not find the attributes mechanical philosophers considered essential to atoms, that is, shape, size and motion:

> by Atoms the Assertors of them understand not indivisible or Mathematicall points which are so void of quantity that the subtle rasor of Imagination it selfe cannot dissect them but minima Naturalia or the smallest particles of bodyes which they call Atomes not because they cannot be suppos'd to be divided into yet smaller parts (for they allow them both quantity & figure as wee shall see anon) but because tho they may be further divided by Imagination yet they cannot by Nature, which not being able in her resolutions of Naturall bodyes to proceed ad infinitum must necessarily stop somewhere & have some bodyes which shee can possibly noe further subdivide & which therefore may be justly termed Atomes.[38]

What Boyle here maintains is: 1. that atoms are not to be confused with mathematical points, but are particles of matter; 2. that although they can be divided by imagination, they are the smallest particles in which natural bodies can be decomposed. The latter property is well elucidated by the following passage:

[37] See above, chapter 1.
[38] BP, xxvi, fols. 162-3.

> In similar bodyes that are really such, for wine milke &c. that some so are not, there being constituted by Atomes is very probable since it is so that their particles are very small & of the same nature with the whole they compose.[39]

Boyle here states that the atoms of homogeneous bodies have the same nature as the whole they compose.[40] The lack of mechanical attributes to atoms indicates that in this manuscript the notion of *minima naturalia* paved the way to Boyle's corpuscular chemistry. This is also confirmed by the experimental proofs Boyle used to support atomism, namely the *reductio ad pristinum statum*:[41]

> thus sylver being dissolv'd in Aqua fortis & that Menstruum so well filter'd that the dissolv'd silver & it will both passe thorough Cap paper all the invisible particles of the Metall which are so small that they hinder not the Diaphaneity of the Menstruum are yet each of them true silver as appeares by precipitating them to the bottome (by a little resolvd salt of Tartar) in the forme of a subtle powder which is easily reducible into the same numericall silver that was at first corroded & so in the mixture of Metalls there is an union per mimima that is Atomes, as if gold & silver be duly melted together each part of the masse has an equall proportion of the respective Metalls, & any part of it being cast into Aqua fortis (which by reason of the virtue wee are now going to ascribe to it is by the French often call'd eau de depart) or water of separation, the Menstruum will corrode and imbibe the Atomes of <the> silver & let those of the gold fall in the forme of powder to the bottome, instances of this Nature might be easily multiplyed if I judg'd them requisite.[42]

Following the *minima naturalia* tradition, Boyle maintains that silver and gold are mixed per *minima* and that the atoms of gold and silver are the smallest parts of these bodies, keeping their (chemical) properties in the

[39] BP, xxvi, fol. 163.

[40] As Newman noticed, Boyle's rejection of the Aristotelian view that milk was homogeneous is indebted to Sennert. See Newman, 'The Alchemical sources' (n. 36), 578.

[41] Cf. C. Meinel, 'Early Seventeenth-Century Atomism. Theory, Epistemology and the Insufficiency of Experiments', *Isis* 79 (1988), 68-103, esp. 92-5. Boyle's example is borrowed from Sennert's *Hypomnemata*. For a comparison of Sennert's with Boyle's text, see Newman, 'The Alchemical Sources' (n. 36), 579-80.

[42] BP, xxvi, fols. 163, 168.

reactions.[43] Like the *reductio ad pristinum statum*, the other proofs (based on the *effluvia* of atoms) Boyle used to support atomism show evidence that at the outset of his scientific career he adopted a qualitative version of corpuscularianism, which in fact does not entirely disappear from the mature version of his theory of matter. As we shall see, later in his career the properties of *minima* are attributed to the compound corpuscles. *Effluvia* of atoms emitted by a number of bodies carry their odours, as well as other qualities:

> In the third place that there are Effluvia or steemes of Atomes issuing out of all bodyes seems not improbable nor that these Effluvia are extreamely subtle that out of odorable bodyes such steemes incessantly issue is of late question'd by few whom the Authority of Ancients has not quite blinded, for that these sents are not bare qualityes but bodyes is evident, both because they may be shut up by close stopping the vessells that containe the bodyes the [sic] flow from because such bodyes actually wast by exhaling, as appeares in chymicall oyles & eminently in Camphir which being very strongly scented tho it be noe liquid body & consequently may seeme indispos'd to <quicke> exhalation, yet left by it selfe in the open aire will exhale all away, & of the subtilty of these Effluvia I observe diverse instances.[44]

Boyle's early attitude towards atomism is also documented by the third essay of *The Usefulnesse*, part I, possibly written in 1651. In this essay, which like the previous ones is aimed at demonstrating that natural philosophy is not incompatible with Christian religion, Boyle placed some restrictions on the heuristic power of atomism:

> For we are yet, for aught I can find, far enough from being able to explicate all the phenomena of nature by any principles whatsoever. And even of the atomical philosophers, whose sect seems to have the

[43] Newman has pointed to the Geberian roots of the view that a mixture takes place through the smallest parts of substances. See 'The Alchemical Sources' (n. 36), and id., 'Boyle's Debt to Corpuscular Alchemy', in *Robert Boyle Reconsidered*, pp. 107-18.

[44] BP, xxvi, fols 169-170. As we have seen (chapter 3), the notion of *effluvia*, as well as that of *minima*, plays a central role in Digby's natural philosophy. Digby appears at the beginning of the manuscript as one of the modern restorers of atomism, together with Gassendi, Magnenus and Descartes (BP, xxvi, fol. 162). Boyle devoted a specific tract to effluvia, i.e., the *Essays of Effluviums* (London, 1673).

> most ingeniously attempted it, some of the eminentest have themselves freely acknowledged to me, their being unable to do it convincingly to others, or so much as satisfactorily to themselves.[45]

Boyle gives two examples of the limits of atomism, i.e., their failure to account for the generation of animals and their inadequate knowledge of the properties of mercury, which Boyle styles the "deluding Proteus". Both these topics will prove rather puzzling to Boyle in the course of his career.

Essays four and five of *The Usefulnese* (which Westfall suggested were written in 1653)[46] display Boyle's increasing anxiety about the atheistic implications of atomism. If Westfall's date for these essays is correct, we can safely infer that Boyle's criticisms of (Epicurean) atomism have been spurred by the revival of Epicureanism of the early 1650s (and in particular by works such as those published by Margaret Cavendish in 1653), which I have dealt with in chapter 3. Boyle's arguments are formulated in six objections to Lucretius's *De Rerum Natura* (lines 416-431 of book 5). The first five deal with the eternity and infinity of atoms – which both Gassendi and Charleton had already criticised. The sixth one deals with the origin of motion in matter, a topic which Boyle regarded as of the utmost importance in polemics against Epicurus, Lucretius and their modern followers. Boyle unambiguously maintains that motion is not essential to matter, but was imposed on it by God. As Boyle put it "matter is no less matter when it rests, then when it is in motion; and we daily see parcels of matter pass from the state of motion to that of rest, and from this to that."[47] God did not create atoms, but a "great mass of lazy matter", which he put into motion. By means of local motion, matter was divided into small particles.[48] In the same text Boyle argues that "matter seems to consist in extension". This

[45] *The Usefulnesse*, *Works*, ii, p. 35.

[46] R.S. Westfall, 'Unpublished Boyle Papers Relating to Scientific Method. I', *Annals of Science* 12 (1956), 63-73 (n. 9 on p. 65).

[47] *The Usefulnesse*, *Works*, ii, p. 42.

[48] *The Usefulnesse*, *Works*, ii, p. 43. Boyle's rejection of the Epicurean view that motion is essential to atoms occurs in his papers against atheism in BP, ii. Boyle maintained that "Eternity, selfe-motion, & c. must necessarily belong to either to God or to matter; and if he will not ascribe them to the Deity, he must doe it to the despicablest Atome." As a consequence, according to Boyle, the Epicurean "make his Atomes so many little gods." BP, ii, fol. 7. Cf. J.J. MacIntosh, 'Robert Boyle on Epicurean Atheism and Atomism', in M.J. Osler (ed.), *Atoms, Pneuma and Tranquillity. Epicurean and Stoic Themes in European Thought* (Cambridge, 1991), pp. 197-219.

favourable attitude towards Descartes's view of matter (though expressed in a hypothetical way) ought not be interpreted as a proof of Boyle's adherence to Descartes's theory of matter, even less of adoption of the Cartesian dualism. It is in fact flatly rejected in *The Origine of Formes and Qualities*.[49] It is likely that in *The Usefulnesse* Boyle adopted it only to oppose the modern atomists' view of self-moving atoms, not because he adhered to Descartes's distinction between *res extensa* and *res cogitans*. Indeed, Boyle's notion of incorporeal spirit (not the chemical or medical ones) is at odds with Descartes's principles.[50] In addition, Boyle's cautious position about the existence of void in nature diverges from Descartes's *a priori* rejection of it.[51] As attested by *The Usefulnesse*, Boyle's main preoccupation when defining his theory of matter was to stress God's absolute power in nature. For Boyle, more than for Gassendi, God's power could not be restricted by any form of natural necessity, including that of creating atoms with specific forms and sizes.[52]

When in *The Origine of Formes and Qualities* (1666) Boyle articulated his theory of matter, he unambiguously distanced himself from both the atomists and Descartes:

> I have forborn to imploy arguments, that are either grounded on, or suppose indivisible corpuscles called Atoms, or any innate motion belonging to them; or that the essence of bodies consists in extension, or that a vacuum is impossible; or that there are such *globuli coelestes*, or such *materia subtilis*, as the Cartesians imploy to explicate most of the phenomena of nature.[53]

Boyle's insistence on the criticism of the theory self-moving atoms may be easily understood if we consider that this view of matter was advocated,

[49] *Works*, iii, p. 7. Cf. A. Pacchi, *Cartesio in Inghilterra* (Bari, 1973), pp. 227-61.

[50] "This immaterial substance [i.e. spirit] either by its own power, or [...] by the assistance of a Spirit of a higher order, such as is God, is able to excite motion in matter and to determine and regulate it." (BP, ii, fol. 61^{v}).

[51] For Descartes's rejection of vacuum, see *Principia*, ii, art. 16-18.

[52] For Boyle's voluntarism, see J.E. McGuire, 'Boyle's Conception of Nature', *Journal of the History of Ideas* 33 (1972), 423-42 and J.W. Wojcik, *Robert Boyle and the Limits of Reason* (Cambridge, 1997), pp. 196-211. For a perceptive assessment of this question see P. Anstey, 'Boyle on Occasionalism: an Unexamined Source', *Journal of the History of Ideas* 60 (1999), 57-81.

[53] *The Origine of Formes, Works*, iii, p. 7.

among others, by Hooke both in the *Micrographia* (1665) and in a letter to Boyle of 1662.[54]

The 'Theorical Part' of *The Origine of Formes* contains Boyle's fully-developed classification of corpuscles, which he had already adumbrated in *The Usefulnesse*.[55] This classification is crucial for the understanding of Boyle's integration of chemistry and the corpuscular theory of matter. As we have seen, the notion of compound corpuscle, or *molecula*, was already adopted by Sennert, Basso and Gassendi.[56] While in these authors the concept of compound corpuscle was marginal, Boyle had frequent recourse to it in the explanation of chemical reactions.

In *The Origine of Formes* we find Boyle's most systematic exposition of his matter theory. He started with the simplest particle, which, as he put it, "though it be mentally, and by divine Omnipotence divisible, yet by reason of its smallness and solidity nature doth scarce actually divide it." Boyle calls these elementary particles *minima* or *prima naturalia*.[57] Unlike the particles we encountered in the manuscript on atomism, the term *minima naturalia* here means corpuscles endowed with purely mechanical properties, namely shape, size and motion. It is remarkable that, unlike Descartes and Gassendi, Boyle does not formulate any hypothesis about the shape of the primitive corpuscles. *Minima naturalia*, by their close union,

[54] R. Hooke, *Micrographia: Or some physiological descriptions of minute bodies* (London, 1665), p. 16. Hooke to Boyle, July, 1662, printed in *Oeuvres Complètes de Christian Huygens* (n. 4), iv, p. 171. Cf. J. Henry, 'Occult Qualities and the Experimental Philosophy: Active Principles in Pre-Newtonian Matter Theory', *History of Science* 24 (1986), 335-381 (esp. pp. 346-347). As I have demonstrated elsewhere, Hooke is the author of a section of Boyle's *Defence of the doctrine touching the spring and the weight of air* (1662), where the spring of air is explained as the effect of the internal motion of atoms, see A. Clericuzio, 'The Mechanical Philosophy and the Spring of Air. New Light on Robert Boyle and Robert Hooke', *Nuncius* 13/1 (1998), 69-75. As J. Henry pointed out, a number of English medical writers adopted the view of active matter. See Francis Glisson, *Tractatus de Natura Substantiae Energetica* (London, 1672).

[55] See *The Usefulnesse*, *Works*, ii, p. 37.

[56] The term *molecula* was employed by Gassendi, see above, pp. 63-6. Cf. H.H. Kubbinga, 'Les premières théories "moléculaires": Isaac Beeckman (1620) et Sébastien Basson (1621)', *Revue d'Histoire des Sciences* 39 (1984), 215-33 and id., 'La théorie moléculaire chez Gassendi', in *Quadricentenaire de la Naissance de Gassendi 1592-1992. Actes du Colloque International Pierre Gassendi. Digne-les-Bains, 18-21 Mai 1992*, 2 vols. (Digne, 1994), ii, pp. 283-302.

[57] *The Origine of Formes*, *Works*, iii, p. 29.

form the primitive concretions or clusters of particles which we can call 'corpuscles of the second order'. According to Boyle, the corpuscles of the second order, which, being too small, are not perceived by our sense organs, are very rarely broken, but remain unchanged in the natural bodies they compose. These corpuscles form clusters of a higher order. Though here Boyle is not explicit about the properties of these corpuscles, we can safely assume that, unlike the *minima* or *prima naturalia* (i.e., corpuscles of the first order), the corpuscles of the second order (and of higher orders) have not just mechanical properties, but also chemical ones. This is what Boyle seems to imply:

> As, not to repeat what we lately mentioned of the undestroyed purging corpuscles of milk, we see that even grosser and more compounded corpuscles may have such a permanent texture: for quicksilver, for instance, may be turned into a red powder for a fusible and malleable body, or a fugitive smoke, and disguised I know not how many other ways, and yet remain true and recoverable mercury.[58]

This point is better illustrated by a passage of Boyle's manuscript on occult qualities:

> such clusters of Corpuscles as are so small as to be below ye Perception of ye Eye yet soe great as may conteyne both sensible qualityes at Last, and specifick propertyes... That likewise such small clusters of Particles as fall not under the sense may reteyne ye whole nature of a Mettall, may be easily evinc'd by diverse Chymicall Experiments.[59]

Boyle's classification of corpuscles according to their different degree of complexity is well elucidated by two fragments of a manuscript. The first reads as follows:

> I think it may be convenient to distinguish the Principles or more primitive, or simple Ingredients of mixt Bodys into three sorts, first Primary Concretions or Coalitions, next, Secondary mixts, and thirdly, decompounded mixts, under which name I comprehend all sorts of

[58] *The Origine of Formes*, *Works*, iii, p. 30

[59] BP, xxii, fol. 120^r, published in M. Boas Hall, 'Boyle's Method of Work: Promoting his Corpuscular Philosophy', *Notes and Records of the Royal Society of London* 41 (1987) pp. 111-143. According to M. Boas Hall, these notes were written before the publication of *The Origine of Formes and Qualities*.

> mixt Bodys, that are of a more compounded Nature, than the Primary, or Secondary ones newly mentioned.[60]

In the other fragment Boyle made a distinction between primary and secondary ingredients of mixed bodies:

> R/r in ye papers about Chymical Principles to make out the distinction betwixt Corpuscles and Ingredients primordial or primary, and secundary [...] & the illustration of the former by Quicksilver, and of the latter by compounded sublimates, Cinabar.[61]

It is not accidental that both of these texts address the problem of the chemical mixture. The aim of Boyle's classification is in fact to make corpuscular philosophy viable to the explanation of chemical reactions. Boyle's use of this classification of corpuscles in chemistry will be dealt with in the next paragraph. What is here important to stress is that, according to Boyle, the properties of compound corpuscles can be entirely different from those of the particles they are made of. They depend on the different textures of simple corpuscles.

Boyle's classification of corpuscles explains what is seemingly a puzzling view of his: that there is no absolute rest in bodies, as even the corpuscles of most solid bodies are in continuous movement – a view which seems to contradict Boyle's rejection of self-moving atoms. This is the topic dealt with by Boyle in a specific work, a short tract entitled *Of Absolute Rest in Bodies*. Here Boyle states that a good deal of experimental evidence shows that the corpuscles of metals, of glass and even of diamonds have some 'intestine' motion.[62] Boyle reported the Epicurean theory of self-moving atoms as a possible explanation of this property of corpuscles, but did not accept it. As often happens with Boyle's works, the reader is left without explanation of the ultimate cause of the proposed phenomenon. With the help of Boyle's classification of corpuscles we can try and solve this puzzle. Though it is clear that motion is not innate to particles of the first order, there are in nature corpuscles which are in fact in continuous movement. These corpuscles can be identified with those forming the primitive concretions (corpuscles of the second order). It is likely that Boyle thought the texture of these corpuscles to be rather unstable and that a

[60] BP, iv, fol. 41^r.

[61] BP, xvii, fol. 154^v.

[62] *Of Absolute Rest in Bodies*, published as appendix to the second edition of *Certain Physiological Essays* (London, 1669), *Works*, i, pp. 443-457.

propensity to motion may follow from this. A passage of *The Origine of Formes* seems to support this explanation:

> it must always happen that the size, and often that the figure of the corpuscle composed by their [i.e., *prima naturalia*] juxta-position and cohesion, will be changed; and not seldom too, the motion either of the one or the other, or both, will receive a new tendency, or be altered as to its velocity or otherwise: and the like will happen when the corpuscles that compose a cluster of particles are disjoined, or any thing of the little mass is broken off.[63]

When Boyle speaks of active corpuscles he evidently has in mind corpuscles of the second order. This is also the case of the seminal principles, which he considered a special kind of corpuscles, namely corpuscles endowed with a formative power. We have already encountered Boyle's notion of seminal principles in the previous paragraph, when analysing his early version of *The Sceptical Chymist*. This text shows that Boyle used this notion (which he reinterpreted in corpuscular terms) to explain the generation of animals, plants and minerals. In *The Usefulnesse* Boyle explains the regular shape of many minerals and of stones, as well as the growth of saltpetre in the earth, as the outcome of seminal principles, that is, of particles which give form to and organise other corpuscles.[64]

What we gather from these texts is that seminal principles are not simple corpuscles, but corpuscles of the second order. In the first part of *The Usefulnesse* we read that some of the first "coalitions or clusters" of corpuscles are "endowed with seminal faculties or properties."[65] This view is restated in *The Origine of Formes,* when Boyle deals with the classification of corpuscles. As he put it, the primary concretion, or corpuscles of the second order, "are, as it were, the seeds or immediate principles of many sorts of natural bodies, as earth, water, salt & c."[66]

While it is beyond doubt that Boyle regarded the seminal principles as special kinds of corpuscles, it is not equally clear what was his view of their origin. In *The Usefulnesse* Boyle maintains that God disposed "that chaos or

[63] *The Origine of Formes*, *Works*, iii, p. 30.

[64] *The Usefulnesse*, *Works*, ii, p. 44.

[65] *The Usefulnesse*, *Works*, ii, p. 37.

[66] *The Origine of Formes*, *Works*, iii, p. 30. This passage is almost a translation from Gassendi's *Syntagma Philosophicum*: "corpuscula quaedam composita subtilissima, ac infra sensus consistenteis, quae sint quasi semina rerum." P. Gassendi, *Opera Omnia*, 4 vols. (Lyons, 1658), i, p. 472a. See also W. Charleton, *Physiologia Epicuro-Gassendo-Charletoniana* (London, 1654), p. 109.

confused heap of numberless atoms into the world [...] especially to connect those atoms into such various seminal textures, upon which most of the more abstruse operations and elaborate productions of nature appear to depend".[67] This clearly indicates that these 'seminal' faculties do not reside in atoms, but are in some compound corpuscles.

Boyle, along with Gassendi, reinterpreted the seminal principles as compound corpuscles endowed with the power of fashioning other particles. Boyle's (and Gassendi's) stress on the seminal principles' faculties and powers, which indeed are not explained in purely mechanical terms, evidently marks a difference from Descartes's and the Cartesians' mechanism. Boyle's own view however differs from Gassendi's in one important point. The former believed that the seminal principles were produced not (as Gassendi maintained) by the spontaneous concourse of atoms, but brought about by God's intervention in nature. According to Boyle, God assembled particles of matter in order to form corpuscles with special textures, so as to produce seminal principles.

The notion of seminal principles is central to Boyle's investigations of the generation of animals and plants. This is the subject of an unpublished essay on spontaneous generation, which seems to have been written before 1657.[68] Boyle unambiguously rejected the view that spontaneous generation takes place in nature. He denied that particles of matter, without the concourse of God, can produce plants or animals. Boyle's view is that seminal principles do not originate from matter only. God "made the Protoplast or the first Individualls of each kind of living Creature, and lodg'd the seminall Principles he thought fit in certain portions of matter." He also maintained that each kind of seminal principle was endowed by God with a specific programme to perform in nature.[69]

[67] *The Usefulnesse*, *Works*, ii, p. 48

[68] The date of composition of the manuscript is discussed in Clericuzio, 'A Redefinition' (n. 35), p. 587.

[69] BP, ii, fol. 141^{v}. Boyle's arguments are directed against the Epicureans' view of generation: "But thô there were cases in which the liveing Creatures that seem to spring of themselves could not be well deriv'd either from the proper thô latent seeds of Genitors of the same kind or from those analogicall seeds that in this Discourse we have called Seminall Principles; yet there will be noe necessity of ascribeing these Productions with Epicurus to blind chance. Since besides the proper and analogicall seeds of Plants & Animals there may be certain things which for want of a fitter name we may call Vicarious seeds, because they may supply the want and performe the part of Seminall Principles more properly soe call'd. For tis

Boyle often referred to seminal principles in his published and unpublished writings devoted to the generation of minerals. His early writings show that Boyle explained the generation of minerals by means of seminal principles. In the first draft of *The Sceptical Chymist* Boyle showed a favourable view of van Helmont's theory of *semina rerum* as the beginnings of minerals.[70] In the first part of *The Usefulnesse* Boyle maintained that, as well as animated bodies, minerals too were generated by "something analogous to seminal Principles, as may appear by their uniform regularity", and by the growth of saltpetre in the earth.[71] Boyle's early views of the generation of minerals have much in common with those of the Paracelsians, of de Clave and of van Helmont.[72] Boyle's theory of the generation of minerals may also be found in a fragment of a discourse in the Boyle Papers on 'Lapidifick Spirit', probably written after the first draft of *The Sceptical Chymist*, as a reference to his water culture experiments seems to testify. In this manuscript Boyle formulated different hypotheses to account for the coagulating power of a 'petrifick spirit'. He suggested that this spirit could operate as a kind of ferment. This, Boyle stated, could be confirmed by the adepts' multiplication. As Boyle put it "there may be even in Metalls a powerfull ferment capable of multiplying it self in quantity as well as in vertue."[73] In this manuscript Boyle also suggested that the

noe way unreasonable to suppose, that as the great Maker of the World is an Omniscient Being, soe when he establish'd the Laws and settled the Course of Nature, he very well knew what Phænomena, must according to such Laws, result from the concourse of such and such causes & circumstances: and soe, that he did both order and foresee that the world being fram'd such as he had made it, a Portion of the universall matter constituteing a liveing Creature of such or such a contrivance, should in such a conjunction of Circumstances appeare with Qualitys or Attributes that should entitle it to such a Denomination." Ibid., fol. 141^{v}.

70 See Boyle, 'Reflections', p. 167.

71 *The Usefulnesse*, *Works*, ii, p. 44.

72 E. de Clave, *Paradoxes ou traittez philosophiques des pierres et pierreries* (Paris, 1635) is quoted by Boyle in *New Experiments and Observations touching Cold*... (London, 1665), *Works*, ii, p. 587. Samuel Hartlib referred to Boyle's possession of a copy of de Clave's work in 'Ephemerides' for 1649, see HP 28/1/32A.

73 BP, xxvii, fol. 301. Whereas in 'Boyle's Alchemical Pursuits' (n. 8) Principe pointed out that not all Boyle's alchemical views are compatible with the principles of the mechanical philosophy, in *The Aspiring Adept. Robert Boyle and his Alchemical Quest* (Princeton, 1998), he interpreted Boyle as a straightforward mechanical philosopher. In my view, Boyle's notion of ferment is not deduced from

'petrific spirit' could operate in a way analogous, but not identical to that of seminal principles strictly understood. Its operations – he wrote – were rather "like that of Runnet when it turns milk into Curds".[74] It seems that Boyle wished to stress the difference between the seminal principles of plants and animals and those of minerals, if such there be. In the *History of Firmness* (1661) Boyle was less confident about the role of seminal principles in the generation of minerals. He did not deny that minerals might be produced by "a plastic Principle implanted by the most wise Creator in certain parcels of matter [...] but the difficulty consists in conceiving, how that internal principle produces its effects."[75] Although he seriously investigated the possible ways in which seminal principles could operate as agents in the generation of metals, he did not exclude other causes. He suggested four causes of the generation of minerals in a manuscript entitled 'Thoughts and Observations about the generation of Mineralls. To be annex'd by way of Additament to the History of Fluidity and Firmness.'[76] The first is that minerals were created with the earth at the beginning of the world – an explanation that Boyle does not endorse, as attested by *The Usefulnesse*.[77] The second way is the "casual coalition of congruous particles." The third way "may be a disposition of Parts amounting to Seminall Principles, or Rudiments, or something able to perform the office of them. And these seminall principles are either lodg'd in Individualls themselves or which is in most cases more probable, in pregnant and prolifick wombs." The fourth way is that performed by subterranean heats and menstruums which settle and harden bodies that were at first fluid. These may be some primitive and uncompounded mineral substances in the

the principles of mechanical philosophy. On the alchemists' ferment see Pietro Bono, *Pretiosa Margarita Novella*, ed. Chiara Crisciani (Florence, 1976) and Martin Ruland, *Lexicon Alchemiae* (Frankfurt, 1612), s.v.

[74] BP, xxvii, fol. 309.

[75] *Certain Physiological Essays*, *Works*, i, p. 434.

[76] BP, xxiv, fols. 1-17. These manuscript notes seem to have been written before 1661, as they are mentioned in *The History of Firmness*, published in *Certain Physiological Essays*, *Works*, i, p. 437.

[77] "To prove, that metalline bodies were not all made at the beginning of the world, but have some of them a power, though slowly, to propagate their nature, when they meet with a disposed matter; you may find many notable testimonies and relations in a little book of *Physico-chymical Questions*, written by Jo. Conradus Gerhardus, a German Doctor." *The Usefulnesse*, *Works*, ii, p. 44. The work referred to is J.C. Gerhardus, *Decas Quaestionum Physico-chemicarum de metallis* (Tübingen, 1641).

bowels of the earth which produce those minerals we are familiar with.[78] The numerous, although fragmentary, manuscript notes which Boyle wrote on this topic show that he was certainly far from relying on strictly mechanical explanations of the generation of metals and minerals, i.e. on the motion of particles of inert matter, but was exploring alternative ones. This is confirmed by what he wrote in the *Essay about the Origin and Virtue of Gems* (1672). In this work Boyle argued that the matter of gems was initially fluid and then hardened into various geometrical forms, according to the figure of the rock cavities it was contained in. Although he was very cautious in asserting a positive theory of the origin of gems, Boyle does not deny that "a seminal and plastick power" is contained in the mineral juices.[79]

CORPUSCLES AND CHEMICAL PRINCIPLES

Thomas Kuhn's statement that Boyle's mechanical chemistry "is incompatible with belief in the existence of enduring elements" has been widely accepted by historians of chemistry.[80] It is beyond any doubt that Boyle rejected the Paracelsian theory that all substances are formed of the same number (three or five) principles – which can be obtained by fire analysis. What Boyle rejected was the Aristotelian, the Paracelsian, as well as the Helmontian claim that all natural bodies are made of the same 'elementary' substances – three or five (the chemical principles), four (the Aristotelian elements) or one (van Helmont's water). This did not mean that he denied the existence of simple and homogeneous substances, but he rejected the chemists' claim that the substance they obtain are simple and homogeneous. He also criticised was the chemists' current techniques of obtaining them from mixed bodies. The fact that Boyle did not pronounce his own opinion about the number of homogeneous substances does not mean that he ruled out their existence. There is at least one substance, which he regarded as simple and homogenous, mercury.

[78] BP, xxiv, fols 1-2, 6.

[79] *Works*, iii, pp. 518, 529.

[80] Kuhn (n. 2), p. 26. This view was accepted by M. Boas: "Where he [Boyle] went wrong is in deciding that there was nothing between corpuscles and compounds, that is, in failing to evolve a totally new concept of element." (Boas, *Robert Boyle*, p. 97). Kuhn's view is also adopted by U. Klein, *Verbindung und Affinität. Die Grundlegung der neutzeitlichen Chemie an der Wende vom 17. Zum 18. Jahrhundert* (Basle-Boston-Berlin, 1994), p. 84.

After Kuhn, historians maintained that the existence of chemical principles is incompatible with corpuscular philosophy. Logical and historical arguments can demonstrate the flaws of this view. It is in fact possible to assert that all bodies are composed of corpuscles endowed with shape, size and motion and, at the same time, to consider the products of chemical analysis (by fire) as the ultimate ingredients of mixed bodies, as well as homogeneous substances. As we shall see in chapter 6, a number of French and Dutch chemists did not reject chemical principles, they reinterpreted them in corpuscular terms. In fact one can conceive of the products of chemical analysis as being real principles and explain their nature and properties in terms of matter and motion.

Boyle's rejection of the chemists' theory of principles is mainly based on experimental grounds. However, he did not peremptorily deny that simple and homogeneous substances can be found – provided that new and more sophisticated analyses are adopted. The fact that Boyle did not provide us with a list of simple substances is due to two main reasons: first, his main target was to disprove the chemists' arguments for their existence and number; second, the lack of empirical evidence made him very cautious on this subject. However, on occasions he spoke of mercury as a simple and homogeneous body. It is however apparent that Boyle did not regard the existence of simple and homogeneous substances as incompatible with the principles of corpuscular philosophy. We can understand Boyle's position on chemical principles if we turn our attention to the classification of corpuscles we have investigated in the previous paragraph.

In *Discourse of the Imperfection of the Chemists' Doctrine of Qualities* Boyle made a clear distinction between the first blocks of matter and the compound corpuscles. The latter have chemical, not just physical properties. On the grounds of this distinction, he considers the chemical principles as primary concretions of corpuscles:

> The chemist's salt, sulphur, and mercury themselves are not the first and most simple principles of bodies, but rather primary concretions of corpuscles, or particles more simple than they, as being endowed with the first, or most radical, (if I may so speak) and most catholic affections of simple bodies, namely, bulk, shape and motion, or rest; by the different conventions or coalitions of which minutest portions

of matter are made those differing concretions that chemists name salt, sulphur, and mercury.[81]

As we shall see, for Boyle the three principles did not have the same status. It would seem that, unlike salt and sulphur, mercury, in his view, could be regarded as a simple and homogeneous substance.

The discussion of the chemical principles is the main subject of *The Sceptical Chymist* and of the appendix (*Producibleness of Chemical Principles*) published in 1680. In *The Sceptical Chymist* Boyle set out "to question the very way of probation employed by Peripatetics and Chymists, to evince the being and number of the elements."[82] Following van Helmont, he questioned the current use of fire analysis and suggested different methods of chemical analysis. As we have seen, Boyle's interest in the *Alkahest* and in more powerful solvents was closely linked with his dissatisfaction with the chemists' analysis. Boyle's adoption of van Helmont's arguments against the Paracelsians' principles did not prevent him from questioning van Helmont's water theory. Boyle stated that water extracted from various bodies still contains particles of other substances, though they are not perceptible to the senses.[83]

The fact that after the publication of *The Sceptical Chymist*, a number of chemists (in England as well as on the Continent) still adopted the Paracelsian doctrine of principles prompted Boyle to return to this subject in other works.[84] Between 1661 and 1666, as we read in the very end of *A Chymical Paradox* (1682), Boyle wrote a discourse to examine "this grand physico-chymical Problem, whether we ought to admit any other elements

[81] *Discourse of the Imperfection of the Chemists's Doctrine of Qualities*, published with *Experiments, Notes, & c. about the mechanical origine or production of divers particular qualities* (London, 1675), *Works*, iv, p. 281. M. Boas has recognised the importance of Boyle's hierarchy of corpuscles: "Boyle had a fairly clear concept of something approaching a chemical molecule; what he conspicuously lacked was any understanding of a modern atom. That is, he went from his *prima naturalia*, which had physical characteristics, but no apparent chemical ones, to the chemical corpuscles which take part in chemical reactions, persist in solution, sublime, and so on, and which obviously may differ in complexity." (Boas, *Robert Boyle*, p. 100).

[82] *The Sceptical Chymist*, *Works*, i, p. 459

[83] *The Sceptical Chymist*, *Works*, i, p. 574.

[84] On the impact of *The Sceptical Chymist* and its relationship to the seventeenth-century chemical textbooks, see Clericuzio, 'Carneade and the Chemists', in *Robert Boyle Reconsidered*, pp. 79-90.

or hypostatical principles at all, even so much as one of the Bodies which are commonly called mixt."[85] The second edition of *The Sceptical Chymist* (1680) came out with an appendix (*The Producibleness of Chemical Principles*) specifically devoted to questioning the chemical doctrine of principles. Here Boyle deals with each of the five principles separately.

The section of *The Producibleness* devoted to mercury testifies to his perplexities about its status.[86] Boyle did not peremptorily deny that mercury was a simple and homogeneous substance. However, he did not decide whether mercury was produced or extracted. He seems to maintain that some mercury is contained in metals, though nobody had extracted it yet. He put forward the hypothesis that the 'metalline' mercury may be considered as a magistery, namely, "a new body produced from some other body without separation of parts."[87]

Boyle's arguments for the existence of mercury in metals are particularly strong: the easy amalgamation of ordinary mercury with metals can be considered as evidence of a kind of 'cognation' between ordinary mercury and the one contained in metals. In addition, he suggests, mercury seems to be the principal cause of metals' gravity:

> The gravity of a metal cannot reasonably be supposed to proceed from the whole body of the metal, but only from some one ingredient heavier in specie than the rest, and than the metal itself. And this ingredient or principle can be no other than the most ponderous body, mercury.[88]

[85] *A Chymical Paradox, Works*, iv, p. 505. Boyle explicitly refers to Thomas Willis's theory of principles: "I do not take chymical principles in the strictest sense of that term, wherein it is confined to salt, sulphur, and mercury; but in the larger acception, wherein the learned Doctor Willis, and divers other chymists (that are not all his juniors) employ it." Ibid., iv, p. 503.

[86] For Boyle's debts to the alchemical 'mercurialist' theories, see Principe, *The Aspiring Adept* (n. 73), pp. 153-80.

[87] Principe, *The Aspiring Adept* (n. 73), p. 54. Boyle defines Magistery as follows: "it is a preparation, whereby there is not an analysis made of the body assigned, nor an extraction of this or that principle, but the whole, or very near the whole body, by the help of some additament greater or less, is turned into a body of another kind." *The Sceptical Chymist*, *Works*, i, p. 637. In Martin Ruland's *Lexicon Alchemiae* (Frankfurt, 1612), *magisterium* is described as "a Chemical state in which matter is developed and exalted by the separation of its external impurities."

[88] *The Sceptical Chymist*, *Works*, i, p. 639. Metals, according to Boyle and to his contemporary chemists, were no homogeneous substances.

In a manuscript in the Boyle Papers we find a more positive statement about mercury. Boyle asserts that the corpuscles of mercury are primary concretions as well as the primary ingredients of some mixed bodies.[89] This statement brings us back to Boyle's classification of corpuscles, which is also dealt with in *The Sceptical Chymist*. Here Boyle distinguishes two kinds of compound corpuscles, according to their textures. One is formed of particles which "are so minute", and adhere so tightly, that it remains unchanged in the mixed body. The other one has a less close texture and is unstable. Consequently, particles of different bodies can separate their constituent particles, which can form a different cluster.[90] As Boyle put it:

> If you allow of the discourse I lately made you, touching the primary associations of the small particles of matter, you will scarce think it improbable, that of such elementary corpuscles there may be more sorts than either three, or four, or five. And if you will grant, what will scarce be denied, that corpuscles of a compounded nature may, in all the wonted examples of chymists, pass for elementary, I see not why you should think it impossible, that aqua fortis, or aqua regis, will make a separation of colliquated silver and gold, though the fire cannot: so there may be some agent found out so subtile and so powerful, at least in respect of those particular compounded corpuscles, as to be able to resolve them into those more simple ones, whereof they consist, and consequently increase the number of distinct substances, whereunto the mixed body has been hitherto resoluble.[91]

As we have seen, the very existence of compound corpuscles endowed with chemical, not purely mechanical properties, makes it possible to assert that in nature exist some homogeneous substances. Though Boyle never assessed what their number and nature are, there is, as we have seen, at least one substance (mercury) which deserves this status.

Boyle's hierarchy of corpuscles has an additional consequence for the theory of chemical composition, that a mixture can be made of corpuscles of different levels of complexity, a view which was to be developed by Joachim Becher. In *The Sceptical Chymist* Boyle articulates this view along the following lines:

> And indeed, as I have formerly also observed, it does not at all appear, that all mixtures must be of elementary bodies; but it seems far more

[89] BP, xvii, fol. 154^{v}.

[90] *The Sceptical Chymist, Works,* i, p. 506.

[91] *The Sceptical Chymist, Works*, i, p. 515.

> probable, that there are divers sorts of compound bodies, even in regard of all or some of their ingredients, considered antecedently to their mixture. For though some seem to be made up by the immediate coalitions of the elements, or principles themselves, and therefore may be called *prima mista*, or *mista primaria*; yet it seems that many other bodies are mingled (if I may so speak) at the second hand, their immediate ingredients being not elementary, but these primary mixts newly spoken of; and from divers of these secondary sorts of mixt may result, by a further composition, a third sort, and so onwards. Nor it is improbable, that some bodies are made up of mixt bodies, not all of the same order, but of several; as (for instance) a concrete may consist of ingredients, whereof the one may have been a primary, the other a secondary mixt body.[92]

One of Boyle's aims was to revise the current classification of chemical substances and to establish more sophisticated criteria to identify different chemical 'families'.[93] As far as salts are concerned, in *The Producibleness of Chymical Principles* Boyle stated that there are three distinct kinds of salt: acid salts (vinegar and spirit of salt), volatile salts (salt of hartshorn, salt of urine, and of blood), and alkalis or lixiviate salts (salt of tartar and potash).[94] The same he proposed for spirits – a term which was used to denote a variety of different substances.[95]

With the term spirit chemists included a variety of substances of different – and often contrary – natures. Under this term chemists included acid substances (as spirit of nitre, spirit of salt, and vinegar), alkaline substances (as spirit of urine), and a third kind of spirits, neither acid nor alkaline. These were very subtle and penetrant, which he called vinous or inflammable spirits.[96]

Boyle's criticism of the traditional chemical classification was also directed against the acid/alkali theory. His objections were threefold: 1. this

[92] *The Sceptical Chymist, Works*, i, pp. 524-525.

[93] Cf. Boas, *Robert Boyle*, pp. 116-122.

[94] *The Producibleness of Chymical Principles*, *Works*, i, p. 596.

[95] "As for what the chymists call spirits, they apply the name to so many differing things, that this various and ambiguous use of the word seems to me no mean proof, that they have no clear and settled notion of the thing." *The Producibleness of Chymical Principles, Works,* i, p. 609. The notion of spirit as the active substance was still central in the chemical books published in the second half of the seventeenth century, see for instance N. Lefebvre's *Course* and T. Willis's *De Fermentatione*, discussed respectively in ch. 6 and ch. 3.

[96] *The Producibleness of Chymical Principles, Works,* i, p. 609.

theory is too comprehensive, as it assumes that all substances are either acid or alkaline. Even in its weaker form, namely that this division applies to salts only, it fails to recognise that there are salts which are neither alkaline nor acid. In addition, it fails to explain many other properties of bodies, like for instance the physical ones, such as malleability, elasticity, magnetism and electricity. 2. It is arbitrary, as salts can be classified according to other properties. Salts commonly classified as alkaline can differ in many respects from other alkaline bodies. Similarly, acids differ from one another in many respects. But acid and alkaline substances can also share a number of properties.[97]

3. The theory is vague and superficial. The way to identify these two classes is still uncertain, being based either on their taste or on the very fact that effervescence is produced when one reacts with the other. Boyle believes that taste is not adequate for this purpose. Furthermore, it is arbitrary to assume – as the advocates of this doctrine do – that when one body dissolves another one, the solvent must be an acid and the dissolved an alkali. Finally, the supposed hostility between these two principles is both an unintelligible and an incorrect notion, as it implies that inanimate beings have attributes which are proper to intelligent ones. Notwithstanding the mistakes and weakness of the acid/alkali theory, Boyle did not deny that, as these differences can be met in a great number of and variety of bodies, "the consideration of them may frequently enough be of good use (especially to Spagyrists, and physicians, when they are conversant about the secondary and, if I may so call them, chemical causes and operations of divers mixed bodies)."[98]

This statement brings us to consider the relationship between mechanical philosophy and chemistry and to assess whether Boyle's chemistry was entirely subordinated to mechanical philosophy or it had its own independent status.

CHEMISTRY AND MECHANICAL PHILOSOPHY

Boyle often repeated that mechanical philosophy was preferable to other theories of matter because it is based on principles which are simpler, more

[97] *Reflections upon the Hypothesis of Alcali and Acidum* (London, 1675), *Works*, iv, p. 286. Boyle's criticisms of the acid/alkali theory were adopted by his assistant and FRS Frederick Slare, see M. Boas Hall, 'Frederick Slare, FRS (1648-1727)', *Notes and Records of the Royal Society of London* 46 (1992), 23-41.

[98] *Reflections upon the Hypothesis of Alcali and Acidum*, *Works*, iv, p. 284.

fundamental and more general than those of both Aristotelians and Paracelsians. Accordingly, it can explain a wider range of phenomena than the other two theories and does not have recourse to such entities as substantial forms, occult qualities, the universal spirit and the *Anima Mundi*.[99] The mechanical philosophy, in Boyle's view, stresses God's creation of the world and his Providence in maintaining it, and helps to dispense with intermediate agents between God and the physical world, which Boyle deemed as both superfluous and dangerous to the Christian religion.[100]

On the basis of these statements, historians have tended to interpret Boyle's science as the fusion of mechanical theory of matter with experimentalism. If restricted to some specific areas of Boyle's science, namely, his study of the spring of air, of hydrostatics and his explanations of electricity, the view of his theories as ultimately mechanical is correct - provided that one does not conceive Boyle's matter theory as *strictly* mechanical. If applied to other fields of Boyle's science, particularly to chemistry and medicine, this interpretation would be wrong.

Nonetheless, when Boyle had recourse to mechanical explanations (in hydrostatics, in magnetism, etc.), he often refrained from reducing phenomena to the shape and size of the simple particles. It is apparent that on occasions he tentatively suggested that crystal shape might indicate the form of invisible corpuscles. Nevertheless, unlike Descartes and the atomists, he did not found his theories on the precise shapes and sizes of the corpuscles. Boyle made this point clear in *The Usefulnesse*:

> it is one thing to be able to shew it possible, for such and such effects to proceed from the various magnitudes, shapes, motions, and concrections of atoms; and another thing to be able to declare what

[99] *Of the Excellency and Grounds of the Mechanical Hypothesis, Works*, iv, pp. 68-78. Boyle's criticisms are also directed against the Helmontian notions of *Archeus*, *Blas* and Gas (p. 69). This criticism does not imply that Boyle saw the Helmontian doctrines not reconcilable with the corpuscular philosophy. As J.W. Wojcik pointed out, in the *Excellency and Grounds of the Mechanical Philosophy* "Boyle believed that a natural philosopher could both accept the corpuscular hypothesis *and* embrace any of a number of a number of other hypotheses, that were consistent with it." Wojcik, *Robert Boyle* (n. 52), p. 181.

[100] See *A Free Enquiry into the Vulgarly Receiv'd Notion of Nature* (London, 1686). Cf. J.E. McGuire, 'Boyle's conception of Nature' (n. 52) and M. Hunter and E.B. Davis, 'The making of Robert Boyle's *Free Enquiry into the Vulgarly Receiv'd Notion of Nature* (1686)', *Early Science and Medicine* 1/2 (1996), 204-71.

> precise, and determinate figures, sizes, and motions of atoms, will suffice to make out the proposed phaenomena.[101]

Besides voicing reservations about the possibility of precisely establishing the primary affections of corpuscles, Boyle put special emphasis on texture - much more than on the shape and sizes of corpuscles. Rather than reducing a phenomenon or a quality to a given shape and size of the particles, Boyle focussed on the texture of corpuscles, namely, on the different kinds of aggregates of particles. This is well illustrated by his views of cold as compared with Gassendi's. The latter saw cold as a true and positive quality and explained it by means of specific atoms ('frigorific atoms'), having the shape of cubes or pyramids and being not easily moved.[102] For Boyle cold was a negative, not a positive quality. Cold is produced by some change in the texture of bodies. His experiments related to cold confirm that Boyle did not have recourse to specific shapes or to the sizes of particles. Boyle mixed sal ammoniac with water and found that the resulting mixture was very cold. The same result occurred when he previously warmed the ingredients. Then he showed that, whereas saltpetre mixed with water produces cold, it produces heat when mixed with oil of vitriol. Then he mixed water, sal ammoniac and oil of vitriol, finding that they produced not cold, but heat. The conclusion was that cold was not produced by one single substance, but by the interaction of different substances. Unlike Gassendi, Boyle did not believe that cold is produced by means of the shape of the fundamental corpuscles. He preferred to refer cold (and heat) to the changes of different textures of the mixtures.[103]

It is apparent that for Boyle physical and chemical properties of bodies are 'relational'. They emerge from the constant interactions of different kinds of corpuscles. Natural phenomena are the outcome of the mutual actions of different kinds of corpuscles being disposed to act on each other. Boyle explained this concept by having recourse to the 'key and the lock' model. As John Henry pointed out, the size and shape of the key is relevant, but is not adequate to explain its power to open the lock. The power of the key derives from the relationship of its shape and size to those of the lock. Boyle adopted this model to explain a number of powers to be found in

[101] *The Usefulnesse*, *Works*, ii, p. 45.

[102] Gassendi, *Opera* (n. 66), i, pp. 398-9.

[103] *Mechanical Origin of Heat and Cold*, *Works*, iv, pp. 237-42.

natural bodies. Corpuscles can produce a given quality only if they act on bodies having a texture which is fit to be acted upon by them.[104]

Boyle's restrictions to the use of mechanical principles are apparent both in chemistry (and mineralogy) and medicine. In the course of his scientific investigations Boyle did not pursue the reduction of chemical phenomena to the more general affections of matter; rather, he had recourse to subordinate theories. He adopted explanations based on the chemical, not on the mechanical affections of particles. The same occurs in his research on medical and biological subjects (that is, the generation of living bodies). Boyle's extensive use of chemistry in medicine, which we have already seen, and his adoption of the seminal principles show that he was reluctant to embrace a reductionist position in medicine.[105] Boyle himself elucidated his position in some scattered remarks (in both published and in unpublished writings) on the scientific method and the different degrees of scientific explanations. "There are oftentimes – he wrote in *Certain Physiological Essays* (1661) – so many subordinate causes between particular effects and the most general causes of things, that there is left a large field, wherein to exercise mens industry and reason."[106] Boyle evidently set out to write an essay on this subject which, as we read in his papers, was lost before being completed.[107] The extant fragments show that Boyle's arguments were directed against those Epicureans and Cartesians "that pretend to explicate every particular Phaenomenon by deducing it from the Mechanicall affections of Atomes or insensible particles."[108] As he put it:

> [These philosophers] are so charm'd with ye clearness & pleasure of Theorys & explications, yt are deriv'd immediately from methaphisical and mathematical notions and & theorems; yt they oftentimes give forced and unnatural accounts of things, rather than not to be thought to have deriv'd them immediately from these highest principles. And, wch is much worse, they despise, & perhaps too condemn or censure all yt knowledge of the works of nature yt Physicians, Chymists & others pretend to, because they cannot be clearly deduc'd from the Attoms, or ye Catholick Laws of motion.

[104] J. Henry, 'Boyle and cosmical qualities', *Robert Boyle Reconsidered*, pp. 119-38.

[105] For the meaning of reductionism, see above, p. 7.

[106] *Certain Physiological Essays*, p. 21, *Works*, i, p. 309.

[107] Royal Society MS 185, fols 31^r-35^v.

[108] BP, viii, fols. 165-166, 169-170, 184-187. Quotation from fol. 166^r.

> The practice of these virtuosi is like, in my opinion, to prove so great an impediment to ye advancement of real learning, yt I think it may be a service to many of its cultivators, to free them from so great a descouragement.[109]

Boyle's conclusions are explicitly anti-reductionist: "The most useful notions we have, both in physics, mechanicks, Chymistry, & the medicinal art, are not deriv'd from ye first principles, but from intermediate Theorys, notions and rules."[110]

If we move from the methodological statements to Boyle's actual chemical and medical works, we find that he made little use of mechanical theories. He explains his frequent references to corpuscles endowed with chemical properties (i.e., compound corpuscles of different degrees of complexity) as follows:

> When I speak of the corpuscles, or minute parts of a body, I mean only such corpuscles, whether of a simple, compounded, or decompounded nature, as have the particles they consist of so firmly united, that they will not be totally disjoined, or dissipated, by that degree of fire or heat, wherein the matter is said to be volatile, or to be fixed. But these combined particles will, in their aggregate, either ascend, or continue unraised *per modum unius*, (as they speak) or as one entire corpuscle. As in a corpuscle of sal armoniac, whether it be a natural or factitious thing, or whether it be perfectly similar, or compounded of differing parts, I look upon the entire corpuscle, as a volatile portion of matter; and so I do on a corpuscle of sulphur, though experience shows, when it is kindled, that it has a great store of acid salt in it, but which is not extracted by bare sublimation.[111]

The *Essay on Nitre*, which aimed at promoting "a good intelligence betwixt the Corpuscular Philosophers and the Chymists", elucidates the status of chemistry in Boyle's science. Boyle's discussion with Spinoza (via Oldenburg) provides good cxamples of two different versions of the corpuscular philosophy. Spinoza's strictly mechanical and reductionist, the other (Boyle's) giving chemical theories an independent status.[112] Boyle

[109] BP, viii, fol 184^{r-v}.

[110] Royal Society, MS 185, fol. 31.

[111] *Experiments, Notes, & c. about the Mechanical Origine or Production of divers particular Qualities* (Oxford, 1675), *Works*, iv, pp. 293-4. See Boas, *Robert Boyle*, p. 101.

[112] On the controversy between Spinoza and Oldenburg/Boyle see R. McKeon, *The Philosophy of Spinoza: The Unity of his Thought* (London, 1928), pp. 137-52;

reported that, by distillation, he had divided nitre into spirit of nitre and fixed nitre, then, by recombining them, he had obtained the same nitre with very little loss of weight. Although in Section XII of the *Essay* Boyle argued that the experiment proved "that motion, figure, and disposition of parts, and such like primary and mechanical affections (if I may so call them) of matter, may suffice to produce those more secondary affections of bodies, which are wont to be called sensible qualities", yet his explanation of this experiment was not grounded on the primary and mechanical affections of particles.[113] His interpretation of the 'redintegration' of nitre was based on the consideration that nitre was a compound body and that the parts into which it was analysed were not "the volatile fixed parts of that concrete" but two distinct substances of different nature, which are obtained from nitre by altering its texture. From this alteration it follows that corpuscles of spirit of nitre and those of fixed nitre "are enabled to disband from concrete and associate themselves with those of their own nature" and acquire a high degree of activity.[114] It is important here to point out that Boyle bestowed upon each of two components – and nitre itself – distinctive chemical properties. Spirit of nitre is an acid spirit, "a kind of *Acetum Minerale*", fixed nitre "is of an Alkalizate nature and participates qualities belonging generally to lixiviate salts", finally, nitre – Boyle continues – "is a peculiar sort of salt, discriminated by distinct properties both from those salts, that are eminently acid, and from those, that are properly alkalizate." The different natures of the two ingredients of nitre are recognized on the basis of the different chemical effects produced by their operations on the same bodies: those minerals that spirit of nitre dissolves, fixed nitre precipitates, those sulphureous bodies which fixed nitre dissolves, spirit of nitre precipitates.[115] Furthermore, they are identified by their tastes and colours. It is noticeable that in the *Essay* Boyle did not attempt to deduce these sensible qualities from the shape or size of the corpuscles of the substances

H. Daudin, 'Spinoza et la science expériméntale: sa discussion de l'expérience de Boyle', *Revue d'Histoire des Sciences et de leurs Applications* 2 (1949), 179-90; A.R. Hall and Marie Boas Hall, 'Philosophy and Natural Philosophy: Boyle and Spinoza', *Mélanges Alexandre Koyré,* edited by R. Taton and F. Braudel, 2 vols. (Paris, 1964), ii, pp. 241-56.

[113] *Certain Physiological Essays*, *Works*, i, p. 364.

[114] Ibid, *Works*, i, p. 369

[115] Ibid., *Works*, i, p. 370.

in question.[116] The process of 'redintegration' of nitre is discussed in terms of chemical qualities and operations of chemical corpuscles. Boyle was in fact aware of the possible objection which could be produced against his account of the experiment, namely, that the 'redintegration' of nitre was the outcome of a mere mechanical association of volatile and fixed parts – an argument which Spinoza did not refrain from using in his criticisms of Boyle's essay. In order to meet such objections, Boyle stated he had obtained the same nitre by combining spirit of nitre with a solution of common potash, or salt of tartar, in the place of fixed nitre. The two substances which he used in the place of fixed nitre were indeed of the same nature as the latter, namely, lixiviate salts.[117] In the artificial nitre the corpuscles of spirit of nitre and those of its fixed salt remain unaltered, but since "they are as it were sheathed up, or wedged in amongst others in the texture of a Concrete" they are not "set at liberty to flock together", and accordingly they cannot display their properties.[118]

It is significant that Boyle's account for the 'redintegration' of nitre, being based on the local motion of corpuscles of acid spirits and of lixiviate salts, provoked the objections of a mechanical philosopher like Spinoza. The two different interpretations of the experiment have been described by the Halls as a typical opposition of a rationalist mechanical philosopher and an empirical one.[119] Indeed, the controversy was also about the relation of chemistry to mechanical philosophy. As a strict mechanist, Spinoza denied that the chemical properties provide information on the nature of bodies. What was important for him was to establish the physical mechanism responsible for a given phenomenon.[120] Boyle was convinced that one cannot understand the properties of a substance (i.e., nitre), by saying that the secondary qualities in question depend on the motion or rest of the corpuscles of 'catholick matter'. As the same mechanisms might produce different effects, the specific properties of a substance can only be understood by means of chemistry. The chemical properties are for Boyle

[116] Boyle refers to the geometrical form of the parts of the 'redintegrated' nitre in order to show that the new substance is nitre. Yet these parts are not the particles of universal matter, but the crystals of nitre. Ibid., *Works*, i, pp. 370-1.

[117] Ibid., *Works*, i, p. 363.

[118] Ibid., *Works*, i, p. 369.

[119] Cf. A.R. Hall and M. Boas Hall (n. 112), p. 242.

[120] For Spinoza's science, see M. Grene and D. Nails (eds.), *Spinoza and the Sciences* (Dordrecht, 1986), in particular D. Savan, 'Spinoza: Scientist and Theorist of Scientific Method', pp. 95-123.

less fundamental than the mechanical ones, yet they must be taken into due consideration.

The Dutch philosopher's first series of objections to Boyle's experiment were contained in a letter to Henry Oldenburg of April 1662.[121] Spinoza's attack was directed to the heart of Boyle's arguments: he contended that spirit of nitre and nitre were not different substances, whereas the fixed nitre had nothing to do with the essence of nitre, being just the impurities of nitre. To Spinoza, nitre and the two parts into which Boyle had analysed it differed only in their mechanical properties:

> In order, then, to explain this phenomenon in the simplest possible way, I will suppose no other distinction between spirit of nitre and nitre itself than that which is sufficiently manifest: namely, that the particles of the latter are in a state of rest, while those of the former are swiftly moved with respect to one another.[122]

The fixed salt was conceived by Spinoza as the outcome of a pure mechanical process, that is to say, a modification of the size of the pores where the nitre was housed. The distinct chemical properties of the nitre and of its spirit, on which Boyle had based his own view that they were substances endowed with different natures, were easily explained by Spinoza on purely mechanical lines. For him, when the particles of nitre were in motion, they produced the acid taste which Boyle deemed the main characteristic of the spirit of nitre. According to Spinoza:

> bodies in motion never come into contact with other bodies along their largest surfaces, while bodies at rest touch other bodies upon their largest surfaces. Thus, if particles of nitre are placed on the tongue when they are at rest, they will lie upon it on their largest surfaces and in this way they block the pores of the tongue, which causes cold. Add to this the fact that the saliva cannot dissolve nitre into such small particles as fire does. But if these particles be placed upon the tongue when they are in excited motion they will touch it with their sharply pointed surfaces and will penetrate into its pores; and the more vigorous their motion, the more sharply will they pick the tongue. In the same way a needle will cause different sensations

[121] For details about this letter, which was published in Spinoza, *Opera*, ed. C. Gebhardt (Heidelberg, 1924), vol. iv, pp. 15-36, and in Oldenburg, *Correspondence*, i, pp. 448-70, see Clericuzio, 'Redefinition' (n. 35), p. 576, n. 72.

[122] Oldenburg, *Correspondence*, i, pp. 449, 459, Gebhardt (n. 121), iv, p. 17.

when it touches the toungue with its point and when it lies flat upon it.[123]

Spinoza argued that the reason why nitre was inflammable but its spirit was not, was that in the former the particles were at rest, whereas in the spirit they were in motion.[124] It is worth noting that Spinoza did not deny experiments a role in the discussion. He in fact claimed that Boyle did not produce sufficient experimental evidence to prove that spirit of nitre was not the same as nitre and that the latter could be produced only by the union of the former with a lixiviate salt. He himself devised three experiments to show that what Boyle called spirit of nitre was indeed pure nitre made volatile by fire.

In the first experiment Spinoza made the volatile particles congeal into icicles of pure nitre. In the second one he showed, by filtering nitre several times, that the volatility depended on its purity. In the third experiment Spinoza demonstrated that, by having spirit of nitre penetrate the pores of sand, its particles lost their motions, and therefore they became inflammable.[125]

The contrast between Spinoza and Boyle was not that of rationalist versus the experimental philosopher, since in fact Spinoza never denied the importance of experiments. The contrast was on the role of mechanical explanations in chemistry. While Spinoza thought that he had explained the redintegration of nitre only when he had reduced all chemical properties of nitre and its components to geometrical and mechanical properties, Boyle accounted for the 'redintegration' on the grounds of the chemical properties of corpuscules, and did not make any attempt to deduce them from the mechanical principles.[126]

The status of chemistry is further elucidated by the *History of Colours* (1664). At the beginning of the work Boyle stated that in chemistry colour change is "sometimes the only thing by which the artist regulates his

[123] Oldenburg, *Correspondence*, i, pp. 450, 460, Gebhardt (n. 121), iv, pp. 19-20.

[124] Oldenburg, *Correspondence*, i, pp. 449, 459, Gebhardt (n. 121), iv, pp. 19-20.

[125] As the editors of the Oldenburg Correspondence pointed out, Spinoza "does not seem to admit the fact of chemical combination, but thinks of corpuscles as being like grains of sand or dust, mere heaps; hence he believes that in adding water he would only be washing out impurities." Oldenburg, *Correspondence*, i, p. 468.

[126] I have dealt with the rest of the Oldenburg/Boyle-Spinoza correspondence in 'Redefinition' (n. 35), p. 577.

proceeding."[127] Boyle's investigations on the nature of colours were – as M. Boas Hall has stated – mostly devoted to studying how matter modified light, rather than the properties of light; and they were entirely based on chemical experiments.[128] Boyle rejected the chemists' view that colours are produced by the chemical principles. He also criticised philosophers who try to explain colours by referring them to the shape and size of the atoms. Boyle stressed the importance of the texture of bodies as the cause of their different colours. Yet, when he explained his experiments he had recourse to corpuscles endowed with chemical properties. In one of the experiments which he devised to confirm his theory of colours he showed that two transparent substances (a solution of common sublimate, i.e., mercuric chloride and water) which he then filtered with oil of tartar (i.e., potassium hydrogen tartrate) make an orange mixture. Then, by pouring into it a few drops of oil of vitriol (sulphuric acid), the solution becomes transparent again. Boyle explained these reactions in purely chemical terms:

[127] *Experiments and Considerations touching Colours* (London, 1664), *Works*, i, p. 669 (hereafter *History of Colours*).

[128] M. Boas Hall, 'Introduction' to the facsimile edition of Boyle's *Experiments touching Colours* (New York and London, 1964), pp. xvi-xvii. Boyle's views of the nature of colours is clearly expressed in the end of the third chapter: "By what has hitherto been discoursed, *Pyrophilus*, we may be assisted to judge of that famous controversy which was of old disputed betwixt the Epicureans and other *Atomists* on the one side, and most other *Philosophers* on the other side; the former denying bodies to be coloured in the dark, and the latter making colour to be an inherent quality, as well as figure, hardness, weight, or the like. For though this controversy be revived, and hotly agitated among the moderns, yet I doubt whether it be not in great part a nominal dispute, and therefore let us, according to the doctrine formerly delivered, distinguish the acceptions of the word Colour, and say, that if it be taken in the stricter sense, the *Epicureans* seem to be in the Right, for if Colour be indeed, though not according to them, but light modified, how can we conceive that it can subsist in the dark, that is, where it must be supposed there is no light; but on the other side, if colour be considered as a certain constant disposition of the superficial parts of the object to trouble the light they reflect after such and such a determinate manner, this constant, and if I may so speak, modifying disposition persevering in the object, whether it be shined upon or no, there seems no just reason to deny, but that in this sense, bodies retain their colour as well in the night as day; or, to speak a little otherwise, it may be said, that bodies are potentially coloured in the dark and actually in the light." *(History of Colours, Works,* i, p. 690). Cf. H. Guerlac, 'Can there be Colors in the Dark? Physical Color Theory before Newton', *Journal of the History of Ideas* 47 (1986), 3-20.

> Colour in our case results from the coalition of the mercurial particles with the Saline ones, wherewith they were formerly associated, and with the alcalizate particles of the Salt of Tartar that swim up and down in the oil. Wherefore considering also, that very many of the effects of the lixiviate Liquors, upon the Solutions of other Bodies, may be destroyed by acid menstruums, as I elsewhere more particularly declare, I concluded, that if I chose a very potently acid liquor, which by its incisive power might undo the work of the oil of tartar, and disperse again those particles, which the other had by precipitation associated, into such minute corpuscles as were before singly inconspicuous, they would become inconspicuous again, and consequently leave the liquor as colourless as before the precipitation was made.[129]

Boyle's explanation of the experiment is based on the substitution of compound corpuscles having chemical properties, not on the mechanical properties of corpuscles. Boyle himself stressed the difference between a mechanical explanation and a chemical one. A mechanical explanation, he maintained, could have demonstrated "why the particles of the mercury, of the tartar, and of the acid salts convening together, should make rather an orange colour than a red, or a blue, or a green." It is however significant that Boyle refrained from formulating a mechanical theory to explain this colour change, but was satisfied with the chemical interpretation of the experiment.[130]

In the 'Reflexions on experiment XL' Boyle refers to a compound corpuscle (of sal ammoniac) as an *explanans*:

> spirit of urine seems chiefly to consist (besides the phlegm that helps to make it fluid) of the volatile urinous salt that abounds in the sal armoniac and is set at liberty from the sea-salt wherewith it was formerly associated, and clogged, by the operation of the alkali, that

[129] *History of Colours*, *Works*, i, p. 762.

[130] Ibid., *Works*, i, pp. 762-3: "This, as I said, Pyrophilus, seems to be the chymical reason of this experiment; that is, such a reason, as, supposing the truth of those chymical notions I have elsewhere I hope evinced, may give such an account of the phenomena as chymical notions can supply us with: but I both here and elsewhere make use of this way of speaking, to intimate that I am sufficiently aware of the difference betwixt a chymical explication of a phenomenon, and one that is truly philosophical or mechanical."

> divides the ingredients of sal armoniac, and retains that sea-salt with itself.[131]

In the practical part of *The Origine of Formes and Qualities*, we find that only on one occasion does Boyle account for the experiments by referring to the "catholick affections" of the particles, namely, in his report of the degradation of gold produced by means of his *menstruum peracutum*.[132] It is however noticeable that when Boyle discussed the transmutation of gold into a baser metal, he suggested that the colour, as well as the other distinctive properties of gold, might be produced by "some noble corpuscles", acting as if they were its tincture. As in the previous cases, Boyle refers to corpuscles having chemical properties.[133] All other experiments to be found in the practical part of *The Origine of Formes and Qualities* are interpreted as the outcome of "the recess of some particles and the access of some others", which are in fact compound corpuscles.[134]

In Boyle's works on physiology mechanical theories play only a small part, as most phenomena are explained in chemical terms.[135] This is attested

[131] Ibid., *Works*, i, p. 765. The role of compound corpuscles in Boyle's explanation of this reaction was stressed by Marie Boas: "Sal ammoniac was clearly what Boyle called a composite, what we now call a compound, in which the component parts could be detached; yet it was also a compound whose corpuscles could be made to act as if they were simple corpuscles, remaining intact through a chemical reaction. Other simple compounds often behaved in much the same way." Boas, *Robert Boyle*, p. 171.

[132] On this experiment, see Principe, *The Aspiring Adept* (n. 73), pp. 79-82.

[133] "However the chemists are wont to talk irrationally enough of what they call *tinctura auri* and *anima auri*; yet, in a sober sense, some such thing may be admitted: I say, some such thing, because as on the one hand, I would not countenance their wild fancies about these matters, some of them being as unintelligible as the Peripateticks substantial forms; so, on the other hand, I would not readily deny but that there may be some more noble and subtle corpuscles, that being duly conjoined with the rest of the matter whereof gold consists, may qualify that matter to look yellow, to resist aqua fortis, and to exhibit those other peculiar phaenomena that discriminate gold from silver; and yet these noble parts may either have their texture destroyed by a very piercing menstruum or by a greater congruity with its corpuscles than with those of the remaining part of the gold; may stick more close to the former, and by their means be extricated and drawn away from the latter." *The Origine of Formes and Qualities*, *Works*, iii, pp. 95-96.

[134] See for instance ibid., *Works*, iii, pp. 83-4, 87.

[135] Mechanical arguments are to be found in Boyle's *Medicina Hydrostatica* (London, 1690), *Works*, v, pp. 453-89.

in the *Memoirs for the History of Human Blood* (1684), a work containing the results of Boyle's investigations of the chemical components of human blood. Boyle reported that from the distillation of blood he had obtained oily and phlegmatic parts, as well as a liquor which he identified as the spirit of human blood. Unlike Willis (*De Fermentatione*), Boyle did not consider these substances as simple and homogeneous. The spirit of human blood, he believed, is composed of volatile salt and phlegm.[136] The properties of the spirit of human blood are described in chemical terms. When Boyle refers to corpuscles, these are endowed with chemical and medical properties:

> the spirit of human blood is endowed with divers qualities, that are both active and medicinal. For it mortifies acid salts, which are the causes of several diseases, and, if I mistake not, of some that are not wont to be imputed to them. It is a great resolvent, and, on this score, fit to open obstructions, that produce more than a few diseases.[137]

When dealing with the relationship between the spirit of human blood and air, Boyle recognised that there is "great cognation or affinity between spirit of blood and air." Yet he was reluctant to establish a theory about the use of respiration. He criticised the view, held by John Mayow, that spirit of nitre is the vital component of air. When it is combined with the sulphureous particles of blood, it produces fermentation and vital heat.[138] According to Boyle, spirit of nitre has no such function, since it is acid and corrosive, while the spirit of blood is an alkali.[139]

Boyle's work on blood and respiration show that he shared the Oxford physiologists' programme to use chemistry as the key to investigate

[136] "The spirit of human blood is totally composed of volatile salt and phlegm; if by phlegm we understand, not simple and elementary water, but a liquor, that although it pass among the chymists for phlegm [...] in the strictest acception it is not that; for when the spirit, volatile salt, and oil are separated from it by distillation and sublimation, as far as they are wont to be in chymical preparations of volatile alcalies, the remaining liquor, which passes for phlegm, will yet be impregnated with some particles of oil, and perhaps also with some few volatile salt, that are too minute to be distinguishable by the naked eye." *Memoirs for the History of Human Blood* (1684), *Works*, iv, pp. 620-1.

[137] Ibid. iv, p. 641.

[138] J. Mayow, *Tractatus duo* (Oxford, 1671, 1st edn: 1668), p. 23. See. Frank, *Harvey*, pp. 224-32.

[139] *The General History of Air* (1692), *Works*, v, pp. 627-92.

physiological phenomena. This did not prevent him from criticising the chemical theories of two physiologists, namely Willis and Mayow.

CONCLUSION

Upon a closer investigation, it has become apparent that Boyle's corpuscular philosophy cannot be described as purely mechanical. Such a definition would ignore the complexity of Boyle's theory of matter, which, as we have seen, includes corpuscles endowed with powers, that is, the seminal principles. In addition, Boyle referred to the texture of bodies, rather than to the size and shape of the ultimate corpuscles. When he explained chemical and physiological phenomena, Boyle referred to compound corpuscles endowed with chemical, not purely mechanical properties. Boyle's hierarchy of corpuscles made it possible for him to explain chemical reactions in terms of access or recess of corpuscles of different substances.

A reassessment of Boyle's theory of matter has enabled us to reconsider his view of principles. As we have seen, his theory of matter was not incompatible with the existence of simple and homogeneous substances. Boyle rejected three central theories of Paracelsian chemistry: 1. that all natural bodies are composed of the same substances; 2. that fire analysis could yield the ultimate components of mixed bodies; 3. that the substances that chemists called principles were simple and homogeneous. He did not deny that simple and homogeneous substances could exist. He tentatively suggested that mercury could be regarded as a simple and homogeneous body.

Another important feature of Boyle's chemical work was his constant concern to produce a new chemical classification of substances. He did not deny that chemical properties could be used to establish 'chemical families'. What he opposed was the current classification adopted by chemists, as it was based on a small number of analogies and overlooked the differences.

As a result of these considerations, the role of Boyle in chemistry can no longer be seen as that of providing a subordination of chemistry to the principles of mechanical philosophy. Boyle's actual chemical work shows that chemistry kept an independent status from physics. On these grounds it is now possible to look at the corpuscular chemistry flourishing in Europe in the last decades of the seventeenth century and to assess the impact of Boyle's chemical theories.

CHAPTER 5

CHEMICAL THEORIES OF MATTER IN ENGLAND AFTER 1661

INTRODUCTION

The aim of the present chapter is to outline the relationship between chemistry and corpuscular philosophy in England after the publication of Boyle's *Sceptical Chymist* (1661). Boyle's ideas played a relevant part in chemistry, at least until the beginning of the eighteenth century. It was not Boyle's intention to establish a systematically organised body of chemical theories, so the impact of his chemistry is not easily measurable. The *pars destruens* of Boyle's programme, namely his rejection of the chemists' theory of principles, was not universally accepted, and a number of chemists and physicians did not rule out this doctrine. It is however relevant that in England, as on the Continent, some of Boyle's objections to the spagyrical theories were accepted. A number of chemists redefined the notion of chemical principle as a useful 'working tool'. They came to consider the three (or five) principles not as the ultimate constituents of bodies, but merely as the products of chemical analysis. In addition, Boyle's quest for a better classification of chemical substances motivated more research on salts, notably on alkalised salts. Finally, the fusion of chemistry and corpuscular philosophy became widely accepted, so that in the 1660s and the 1670s a number of Helmontians reinterpreted the main notions of van Helmont's iatrochemistry in corpuscular terms.

IATROCHEMISTRY AT OXFORD

In the third edition of Willis's *De Fermentatione* (1662) the theory of principles is subject to some changes which may be explained as an answer to Boyle's objections. Willis allowed that the chemists' principles were not the ultimate constituents of bodies, yet he restated the view (which Boyle had rejected) that they were contained in all natural bodies.[1] In *De Sanguinis*

[1] In the Preface to *The Sceptical Chymist* Boyle evidently referred to Willis and other English physiologists when he wrote: "For I observe, that of late chymistry begins, as indeed it deserves, to be cultivated by learned men, who before despised

Incalescentia (1670) Willis revised his view of spirit as expounded in *Diatribae duae* (1659). There he had explained vital heat and blood fermentation as a result of the action of spirits. Later (in 1670) Willis felt that the notion of spirit was too vague. The current research on nitre had provided him with new theories about vital heat. As a result, this was no longer interpreted as the outcome of fermentation, but (after Mayow) as the consequence of the chemical reaction of corpuscles of nitre, contained in air, with the sulphurous ones contained in blood.[2]

In Mayow's first work on respiration (*Tractatus duo*, 1668) aerial nitre is deemed to be the source of life. Its particles produce fermentation when they are mixed with the sulphurous ones contained in blood.[3] In *Tractatus quinque* Mayow slightly changed his theory of nitre and dealt with the theory of matter, though not extensively. Mayow, who was certainly aware of Boyle's objections, maintained that the vital component of air was not spirit of nitre, but only the more subtle and volatile part of it, i.e., the nitro-aerial salt.[4]

Mayow rejected the mechanical philosophy and kept a *via media* between the chemists and the mechanical philosophers – a position which his Oxford colleagues shared. His theory of matter was corpuscular, but he did not reject the chemists' theory of principles. He emphasised the

it... whence it is come to pass, that divers chymical notions about matters philosophical are taken for granted and employed, and so adopted by very eminent writers both naturalists and physicians. Now this. I fear, may prove somewhat prejudicial to the advancement of solid philosophy" *Works*, i, p. 459. See Willis, *De Fermentatione*, in T. Willis, *Practice of Physics* (London, 1684), p. 2.

[2] Willis, *Practice of Physics* (n. 1), pp. 21-3.

[3] See Mayow, 'De respiratione', *Tractatus duo* (Oxford, 1668). On Mayow see Partington, ii, pp. 577-613 and Frank, *Harvey*, pp. 224-232. Before Mayow the spirit of nitre was conceived as the source of life by Sendivogius and by R. Bathurst. See H. Guerlac, 'The Poets' nitre. Studies in the chemistry of John Mayow. II', *Isis* 45 (1954), 243-55; A.G. Debus, 'The Paracelsian Aerial Niter', in A.G. Debus, *Chemistry, Alchemy and the New Philosophy* (London, 1987), ch. ix.

[4] According to Mayow, spirit of nitre is composed of two distinct substances, only one of them necessary to life: "But when I had seriously considered the matter, the acid spirit of nitre seemed to be too ponderous and fixed to circulate as a whole through the very thin air. Besides, the nitro-aerial salt, whatever it may be, becomes food for fires, and also passes into the blood of animals by means of respiration. But the acid spirit of nitre, being humid and extremely corrosive, is fitted rather for extinguishing flame and the life of animals, than for sustaining them". *Tractatus quinque*, repr. in John Mayow, *Medico-Physical Works* (London, 1907), p. 8.

importance of local motion of corpuscles, but did not reduce chemical reactions to the mechanical affections of matter. As he put it:

> I do not think we ought to agree with recent philosophers, who believe that fire can be produced by the subtle particles of any kind of matter if they are thrown into violent agitation. In fact, while the Peripatetics formerly assigned a distinct quality for almost every natural operation and multiplied entia unnecessarily, the Neoterics on the other hand maintain that all natural effects result from the same matter, its form and its state of motion or of rest alone being changed, and that consequently any thing whatever may be obtained from any thing. But in truth this new philosophy seems to depart too far from the doctrine of the ancients, and I have thought it better to take an intermediate path. It would certainly be a reasonable supposition that certain particles of matter which are unlike in no other respect than in the form and extremely solid and compact texture of their parts, differ so much that by no natural power can they be changed one into another, and that the elements consist of primary, and in this way peculiar particles.[5]

I do not believe that – as Partington argued – this statement was directed against Boyle. It is very likely that Mayow's main target was Cartesian mechanism.[6] The explanation of chemical reactions by means of compound corpuscles, as we have seen, was in fact central to Boyle's chemistry. However, despite Boyle's criticism, Mayow still adopted the chemical theory of principles.

HELMONTIANISM AND CORPUSCULAR PHILOSOPHY

Helmontian iatrochemistry enjoyed great popularity in the second half of the seventeenth century. It found in George Starkey an influential advocate. Along with van Helmont, he was a radical opponent to Aristotelian philosophy, and stimulated other attacks on the medical establishment. He was followed by Marchamont Nedham, George Thomson and William Simpson.[7] Unlike Starkey, who adopted a corpuscular theory of matter, the

[5] Ibid., pp. 5-6. As most of his contemporaries, Mayow explained the generation of metals by means of the seminal principles.

[6] J.R. Partington, 'The Life and Work of John Mayow' *Isis* 47 (1956), 217-30 and 405-417 (esp. p. 406).

[7] See A. Clericuzio, 'From van Helmont to Robert Boyle. A study of the transmission of chemical and medical theories in seventeenth-century England', *The*

other three Helmontians paid little attention to the matter theory, but focused on chemical and medical themes. It is however noticeable that both Thomson and Simpson endorsed van Helmont's criticism of the Paracelsian doctrine of principles and adopted the theory of water and seeds. Simpson explicitly rejected Willis's theory of principles and referred to *The Sceptical Chymist* to support the Helmontian view.[8] As we shall see, Simpson's subsequent works bear evidence of his adoption of Boyle's corpuscular philosophy. Such a combination of the latter's with van Helmont's doctrines became common in Restoration England.

Since the mid-1660s a number of English chemists, i.e., Daniel Coxe, John Webster, Thomas Sherley, William Simpson and William Bacon interpreted van Helmont's iatrochemical doctrines in corpuscular terms. Besides maintaining that water was the material principle of all bodies, they deemed van Helmont's *semina*, ferments and spirits as corpuscles endowed with a plastic formative power. They also put special emphasis on the operations of the *Alkahest*, which they interpreted (after Boyle and Starkey) in corpuscular terms. Boyle's *Sceptical Chymist* and *The Usefulnesse* were often quoted to support the corpuscular interpretation of Helmontian theories. Coxe was also personally acquainted with Boyle and corresponded with him. As Coxe's chemical views are significantly articulate, I deal with them in the next section.

In 1671 John Webster, who in the Interregnum was among the early advocates of Helmontianism, published *Metallographia,* a work aimed at improving the knowledge of minerals and metals. He draws on a vast amount of sources, and adopts van Helmont's theories in his explanation of the origin of minerals. He asserts that minerals have a "seminary principle to propagate themselves by", and are generated in the bowels of the earth, where they are contained in a liquid form.[9] The transmutation of metals into gold is interpreted as the passage of metals from the state of unripeness to that of maturity. Webster's account for the operation of the 'metalline seed' is not based on the principles of mechanical philosophy, namely, as a mere

British Journal for the History of Science 26 (1993), 303-34, esp. pp. 319-26. On Starkey's medical work see Newman, *Gehennical Fire*, pp. 175-208.

[8] See G. Thomson, *Aimatiasis, or the true way of preserving the bloud* (London, 1670), p. 31. W. Simpson, *Hydrologia Chymica* (London, 1669) pp. 246, 258-9.

[9] J. Webster, *Metallographia, or an History of Metals* (London, 1671), pp. 41-2. On John Webster see A. Clericuzio, 'Alchimie, philosophie corpusculaire et minéralogie dans la *Metallographia* de John Webster', *Revue d'histoire des sciences* 49 (1996), 287-304.

juxtaposition of atoms of inert matter. For him, the aggregation of atoms brings about only "an increase in the bulk". He had recourse to a plastic principle to explain the origin of stones and of minerals:

> in Vegetables there is a seminal spirit, vapour, or steam that transmute and assimilate, and in stones there is a petrifying quality, vapour or steam, that doth turn the matter aggregated, into the nature of this or that stone.[10]

Ignoring Boyle's doubts about van Helmont's account of the generation of metals from seeds, John Webster quoted from *The Sceptical Chymist* to reinforce his own view that metals were generated from seeds and passed from the liquid to the solid state by the action of a "formative principle".[11]

More than Webster's *Metallographia*, Thomas Sherley's *Philosophical Essay* proposed a fusion of Helmontian doctrines with the corpuscular philosophy. Like other Helmontians, Sherley supported the view that water and *semina* are the principles of all natural bodies. His own interpretation of *semina* was unambiguously corpuscular. Sherley's *semina* are particles of matter having a specific programme:

> By seed I understand a fine, subtile substance, (imperceptible by our gross Organs of Sensation) in which God hath impressed a Character of that thing he will have it produce from the Matter it is to work upon: which it doth perform by putting the parts of matter into such a peculiar motion as it is requisite to produce the intended effect.[12]

Sherley insisted on the formative power of seeds – a power which he saw as originating from God, and directing the particles of matter according to specific ends. He also dealt with ferments, which he regarded as an "expansive power" contained in the seed, producing a change in the texture of the body.[13]

Ferments and fermentation are the subject of William Simpson's *Zymologia Physica* (1675). In this work Simpson asserted that corpuscles endowed with mechanical properties could explain fermentation, which in fact is necessarily brought about by some power. Such a power keeps the

[10] Webster, *Metallographia* (n. 9), p. 46.

[11] Ibid., pp. 50, 54-9.

[12] T. Sherley (or Shirley), *A Philosophical Essay* (London, 1672), sig. A6^v-A7^r. For Shirley see A.G. Debus, 'Thomas Sherley's *Philosophical Essay* (1672): Helmontian Mechanism as the basis of a new philosophy', *Ambix* 27 (1980), 124-35.

[13] Sherley, *Philosophical Essay* (n. 12), sig A7^r and pp. 24; 32; 35.

particles of matter in continuous and rapid motion, namely, the Helmontian *semina* – which he saw as corpuscles containing a spiritual formative principle. Simpson's adherence to corpuscularianism was unambiguous, yet he stressed that the motion of corpuscles is directed by plastic principles, which he identified with the Helmontian *semina*.[14]

The use of Boyle's works to legitimise the fusion of Helmontianism with the corpuscular philosophy is particularly apparent in William Bacon's *A Key to Helmont* (1682), where both the Aristotelian and Paracelsian theories of matter are flatly rejected. Water and spirits are for William Bacon the two principles of natural bodies. The spiritual agent operates as a plastic principle, giving the corpuscles of water different textures, thereby producing natural bodies.[15]

CHEMISTRY AT THE ROYAL SOCIETY: DANIEL COXE AND NEHEMIAH GREW

The experimental attitude of the Royal Society did not prevent Fellows from reading and discussing papers dealing with chemical theories. The chemical principles, the status of volatile salts extracted from plants, the acid and alkali doctrines, as well as fermentation and the chemical properties of air were all topics discussed in the Society meetings. What is apparent is that different views of the chemical principles, as well as diverging interpretations of the corpuscular theory of matter, coexisted. Iatrochemistry (and Helmontianism) were also represented (Goddard and Coxe), while Fellows like Hooke and Slare developed chemical investigations devoid of Helmontian undertones. It is interesting to note that alchemical topics were not banished from the Society.[16]

Though the name of Daniel Coxe has come up from time to time in a variety of contexts concerning English science (there is a biographical sketch in the *Dictionary of National Biography*), no-one has yet attempted to study his chemical papers and letters. His extant papers are of considerable interest for the history of chemistry. They deal with topics

[14] W. Simpson, *Zymologia physica, or a brief discourse of fermentation* (London, 1675), sig. A4^r; pp. 4-7. See also id., *Philosophical Dialogues* (London, 1677).

[15] W. Bacon, *A Key to Helmont* (London, 1672) pp. 1-3, 31-2.

[16] See T. Birch, *The History of the Royal Society of London*, 4 vols. (London, 1756-7), vol. i, p. 123 (a paper on the degradation of Gold read by P. Stahl); ii, pp. 97; 105; 107; 113 (paper on transmutation read by Coxe).

which are central to seventeenth-century chemistry, like the principles of mixed bodies, the classification of salts, and their extraction from plants, as well as with the preparation of the universal solvent. Moreover, they shed light on the chemical interests of the Royal Society in the 1660s and 1670s and on the impact of Boyle's chemistry among his contemporaries.

Coxe's interest in natural philosophy started very early, possibly in 1655. According to Frank, he was somehow connected with the Oxford Experimental Club in the late 1650s (c. 1657).[17] He was certainly linked with John Wilkins, since the latter proposed him as candidate Fellow to the Royal Society on 15 March 1665. Coxe was elected on 22 March 1665 and at the meeting of 19 April read his first paper at the Royal Society, i.e., 'Some Inquires on the vegetation of plants'. It addresses, among other topics, spontaneous generation and the composition of seeds. It is noticeable that experiments on spontaneous generation were proposed by Boyle to the Society on 17 May of the same year. In the months following his election, Coxe was very active in the Royal Society. His subsequent contribution was on poisons – a topic which, as we have seen, was part of Boyle's agenda.[18] In 1665 Coxe started his correspondence with Boyle. The first extant letter (undated, but probably written in late August or early September 1665) bears evidence of their collaboration. Coxe informs Boyle of his current chemical work, i.e., the preparation of solvents, transmutations, and the volatilisation of salt of Tartar [potassium carbonate] – a substance that played a prominent part in van Helmont's and Starkey's chemistry. As we shall see, the volatile salt of tartar is, for Coxe, the main ingredient of the *Alkahest*.[19] From Coxe's extant letters (Boyle's to him are all lost), we gather that Coxe's chemical research was encouraged by Boyle.[20] It is also apparent that Boyle sent instructions to Coxe on a number of chemical processes. In a letter to Boyle, possibly written in September 1665, Coxe gave an account of his research, which includes the preparation of volatile

[17] Frank, *Harvey*, pp. 61; 70-1.

[18] See Boyle's manuscript notes in Royal Society MS 1, fols. 74^r-88^v, Oldenburg's letter to Boyle of 10 August 1665 and Boyle to Oldenburg of 12 August 1665, Oldenburg, *Correspondence*, ii, pp. 457-60 and 475-6.

[19] Coxe was not the only Fellow of the Royal Society to adopt Helmontian views. They were also adopted by Jonathan Goddard. For Jonathan Goddard, see Webster, *Instauration*, pp. 55; 57; 79-81.

[20] Coxe to Boyle, Boyle Letters, ii, fols. 76-7. Coxe's letters are included in M. Hunter and A. Clericuzio (eds.), *The Correspondence of Robert Boyle*, 6 vols., forthcoming (vols. ii and iii).

salts from vegetables, the volatilisation of spirit of sea-salt, and of a number of Helmontian recipes – including *elixir proprietatis* and *Aroph* (ferric ammonium chloride).[21] The preparation of solvents was central to Coxe's chemical investigations. Along with Boyle, Coxe maintained that powerful menstruums could decompose bodies into their constituents, which ordinary fire analysis fails to do. Coxe also claimed that with the help of a powerful solvent the transmutation of metals could be achieved. In a letter to Boyle of 19 January 1666 Coxe stated:

> my former, & more recent Experiments having Thoroughly Instructed mee that no Considerable progresse can bee made in disquisitions Concerning any Concrete in nature, especially Mineralls unlesse wee are masters of some Excellent Menstrua, from whose assistance wee may derive many Advantages. Either analyzing the bodies wee operate on & thereby Informe our selves what theire Constituent principles are; & Consequently wee might increase our power over nature whose Products wee might not only neerly imitate, but also perhaps meliorate, & Improve [...] I confesse indeed most of the ordinary analyzers (such are corrosive Acid spiritts & Salts) divide into Integrall only, not Elementary parts. But yet Neverthelesse I am fully satisfied that there are menstrua existent, or att least such may bee procured which will resolve even the most solid fix'd bodies into their Component principles. I am induced to Entertain this persuasion from the Consideration of Mettalls & Menstrua in generall.[22]

Coxe believes that metals are composed of saline, sulphurous, mercurial and earthy parts – substances which he interprets in corpuscular terms. He argues that the subterranean heat reduces these principles to their smallest parts, which then meet together and adhere, forming the molecules of metals. Coxe is one the early chemists to make consistent use of the notion of molecule in chemistry. Following Gassendi, Coxe introduced the notion of molecule (i.e. compound corpuscles) in his theory of matter. For Coxe, molecules – not their component particles – can be decomposed by ordinary menstruums, but their components particles cannot. The size of the component particles and their close texture make them very resistant to the action of ordinary solvents. As Coxe put it:

> Now the Cohesion of the molecula is so loose (by reason of the Comparative greatnesse of theire parts) that many ordinary grosse menstrua may Easily Enough separate them, but the Texture of the

[21] Boyle Letters, ii, fol. 72. On *Aroph*, see Partington, ii, p. 226.
[22] Boyle Letters, ii, fol. 54.

> first Principles being more close, they are not separable but by Some Extreeme Subtle Analyzer which by reason of the minutenesse brisk motion, & Convenient Figuration of itt's parts may disunite them.[23]

Evidently, the preparation of van Helmont's *Alkahest* was the aim of Coxe's chemical work. As I have shown elsewhere (chapter 4), Boyle regarded the *Alkahest* with the utmost interest. Like Boyle, Coxe believed that the universal solvent could perform extraordinary operations, besides providing a good deal of information on the composition of bodies:

> I have often thought with an excesse of pleasure what wonderfull operations wee might performe by the mediation of this Alcahesticall liquor. For if I were master of itt I should expect that itt should furnish mee with Experiments lucriferous, & Luciferous. For discovering to mee the Constituent principles of bodies and what proportion of them went to make up the Compound, I might bee enabled to imitate Nature & Produce Even Gold gems.[24]

Later, Coxe's enthusiasm faded away, and he was no longer sure that what the *Alkahest* yields are the ultimate constituent of bodies:

> But my more mature deliberate thoughts have suggested more doubts to mee then I can easily resolve. I find that itt would bee a hard task to satisfy a scrupulous judicious enquirer that the substances wee obtaine were Praeexistent in that forme before they constituted the Concretes which wee suppose to be the result of their Union.[25]

Notwithstanding his perplexity about the properties of the *Alkahest*, Coxe goes on to give a mechanical explanation of the way the immortal solvent might operate:

> The small parts of bodies cohering only by immediate contact, or Rest, there seem to bee few bodies whose constituent parts are so closely united, but that the minute parts of this menstruu[m] insinuating themselves between dissolves the Cement or Structure which kept the different bodies under the same forme, so that the particles being disjoyned naturally associate with their like, and possesse those places which divers degrees of gravity Levity, Fixidnesse, or Volatility whither assigned by nature or the Menstruu[m]. From these premises I conjecture the Alcahest to be a liquor consisting of small, yet solid parts, vigorously moved; By

[23] Ibid.

[24] Ibid.

[25] Ibid.

> mediation of which subtlety solidity Convenient figure & Agitation itt becomes the Idoneous menstruum of most bodies few Cements in nature being able to Elude itts Analyzing power, although some admitt of a more Easy Solutions then others.[26]

Having thoroughly examined van Helmont's *Ortus Medicinae,* Coxe concludes that Salt of tartar is the main ingredient of the *Alkahest.*

Besides the universal solvent, the other focus of Coxe's chemical work is the properties of salts – a topic which he dealt with extensively in the papers he published in the *Philosophical Transactions* for 1674.

In 1665 Coxe tells Boyle that he has reached the conclusion that Alkalies or fixed salts, obtained by incineration of plants, do not pre-exist in them, but are produced by fire. He maintains that, before the action of fire, the Alkali salt is volatile. Coxe's explanation is corpuscular:

> Vegetables only afford this Alcali Salt [...] now only a Nitrous salt is extracted out of the Earth from which it derives itts fertility and vegetables a great part of theire Nourishment: now this salt dissolved in water and conveied by the ordinary channell into the Plant itt is easy to evince that by a Mechanical necessity itt must in itts Passage & perhaps in itts station be considerably exalted or volatilized by the active Principles with which itt is associated, the Action & Reaction being mutuall: Nay Perhaps by itts sole motion and passage through the straite pores of the vegetable itt may bee attenuated or broken into such small portions or parts that itts motion may bee Prevalent over it's Gravity which is the very Essence of Volatility.[27]

It is noticeable that, for Coxe, the volatilisation of the salt is performed by some active principles – a plastic agent or a seminal principle. Like Boyle, Coxe recognises that the nature of this architectonic principle is rather obscure. He is convinced that one can only make conjectures about its nature and properties. However, Coxe does not refrain from informing Boyle of his thoughts on the subject. Coxe's explanation is based on the analogy with alchemical *opus* and on van Helmont's account of the projection:

> Probably besides shee [Nature] imploys a seminal forme or Architectonick principle wherby these Substances acquire Such a Schematisme from whence this aggregate of particles derives itts denomination. I do not here understand any occult Quality, or

[26] Boyle Letters, ii, fol. 55.
[27] Boyle Letters, ii, fol. 55.

> incorporeall Substance, but either a Constant, determinate degree of heat perhaps not imitable by art. Or Some Substance Exceeding Small in Quantity, but very Powerfull in Energy by whose irradiation matter is disposed to acquire the forme of Gold, silver, or some other minerall according to the nature of the Active Archeus, or Passive Substances the subjects of itts operations. I am confirmed in these Sentiments by the account Helmont gives of his *Pulvis Chrysopoeius*. Hee affirms that a small quantity of the powder of Projection perfectly transmuted a Comparatively vast quantity of Crude metal.[28]

Coxe goes on to express his views of the transmutative agent. He maintains that gold is not the only source of the tincture, which can also be found in "Concretes in nature base and Ignoble in our esteeme & cheap enough which containe more of the Architectonick principle, commonly stiled the Tincture of gold."[29]

It is apparent that Coxe is well informed of Boyle's alchemical investigations – a fact which explains why Boyle chose Coxe (with Dickinson and Locke) for the examination of his chemical papers. In January 1666 Coxe asks Boyle to send him information on the transmutation of metals. He promises he will keep the process secret.[30]

Following van Helmont, Coxe maintains that there are two kinds of 'metallic sulphur', one internal, the other external:

> Sulph. seems to bee of 2 sorts, the one Conspiring with the other principles to Constitute the pure metallic part and is so strictly united with them that not separable but by the mediation of some Alcahesticall menstruu[m]. The other lies Pretty lax or loose and is So plentifull in some mineralls, that when the concrete is exposed to the fire itt carries up the metalline part inveloped therein.[31]

[28] Ibid.

[29] Ibid.

[30] "Communicate somewhat Considerable on this subject, with some hints att least about the volatile Salt of Tartar. The Compendious way of Distilling Mercury: Method of reducing Alcalis into earth: the flux powder which accelerates the fusion of mettalls. If you will bee pleased to intrust mee with these Arcana assure your selfe Sir you shall never have cause to repent that ever you reposed so great s Confidence in mee. For I shall faithfully Conceale them if you oblidge mee to secrecy & industriously improve them as far as my small Ingeny & Experience will permitt." Ibid.

[31] Ibid.

When describing the particles of sulphur, Coxe rather follows Descartes, stating that they are small, branched and smooth. Accordingly, they can easily embrace the particles of salt. Coxe discussed the properties of salts at the Royal Society on 14 March 1666. He gave an account of the different figures of crystals of salts and stated that different salts can be distinguished "by the alteration of the figure caused by the addition or mixture of something else."[32]

Coxe's familiarity with Boyle's chemical work prompted Wilkins (25 April 1666) to ask Coxe to replicate Boyle's experiment of turning water into earth – an experiment described in *The Origine of Formes and Qualities*.[33] Unfortunately, there is no evidence of Coxe's report to the Royal Society. It is however interesting that on 4 June 1666 Coxe reported to the Society that he had successfully tried another Boylean experiment, namely the transmutation of gold into silver, contained in *The Origine of Formes and Qualities*.[34] The Society was evidently interested in the transmutation of gold into silver, as on 18 July 1666 Coxe was asked to bring to the subsequent meeting the white powder made of gold, and the vitriol used to produce gold.[35] On 29 August Coxe in fact produced some of the white powder and was asked to repeat the experiment with a greater quantity of gold (at the Society's expense).[36] In subsequent meetings Coxe informed the Society of his investigations on vegetables, reporting (on 28 May 1668) that he had already written more than one hundred pages on this subject and gave an account of it. The work was never finished, but a summary of it can be found in a letter from Oldenburg to Martin Vogel dated 13 February 1669.[37]

In 1674 Coxe's papers on volatile salts extracted from plants appeared in the *Philosophical Transactions*. Coxe's papers were part of his planned history of vegetables. As Oldenburg stated in a note published at the end of the first paper, the experiments therein discussed were performed in 1666 and then communicated to Boyle. In the first paper Coxe maintains that the volatile spirit extracted from vegetables, after two or three rectifications, becomes a urinous spirit, like spirit of urine or sal ammoniac. Coxe gave a detailed description of its properties: "These volatile spirits and salts

[32] Birch, *History* (n. 16), vol. ii, p. 67.
[33] Ibid., p. 86.
[34] Ibid., p. 97.
[35] Ibid., p. 105.
[36] Ibid., p. 113.
[37] Oldenburg, *Correspondence*, v, pp. 404-7.

[extracted from vegetables] have not only the same sensible properties; but also agree in all known operations with common Urinous spirits and salts; as in the changing Syrup of violet and many other vegetable tincture green; in being Diaphoretic, Diuretic, and De-obstruent: Contrary to Acids, which they do mortifie. They unite with Acids, and thereby become neutral Salts."[38]

The second paper deals with a topic which Coxe had already treated in his correspondence with Boyle, i.e., the alkalizate or fixed salts extracted from the ashes of plants after incineration. Coxe firmly denies that fixed salts pre-existed in vegetables, they are produced, not extracted, by the fire. The salts alkali result from the combination or union of the saline and of the sulphurous principles. As a result of the action of fire, volatile salts combine with sulphur – producing alkalized salts. This is confirmed, for Coxe, by the fact that alkalis can be divided into oil and volatile salt. In the same paper Coxe maintains that the fixed salts extracted from the ashes of plants do not differ from each other. In the third paper, published in the *Philosophical Transactions* for November 1674, Coxe argues that all "volatile salts, being freed from adhering oyles or sulphurs, become forthwith Homogeneal and Uniform."[39] Having established that volatile salts are homogeneous substances, Coxe maintains that volatile salts are the vital component of air. Their origin and properties are described along the following lines:

> [Air] is impregnated with a Volatil Salt, partly sublimed by Subterraneous, and extracted by Celestial, Fires; partly expired from animals during their life; and both from them and Vegetables upon the dissolution or dissociation of their constituent parts in rarefactions and Fermentations.[40]

Daniel Coxe's chemical work shows that Boyle's corpuscular chemistry was behind a large portion of the chemical investigations carried out in the early Royal Society. Coxe, who often had the role of Boyle's *alter ego*, played an important part in the chemical experiments of the Royal Society.

[38] *Philosophical Transactions* 101 (1674), 4-8, esp. p. 8.

[39] Ibid., 171.

[40] Ibid., 172-3. The paper of November 1674 contains some reflexions on the regeneration of plants from their ashes. Coxe observes the similarity of the ashes of fern with the plant and seems to believe that their regeneration is possible. There is another set of papers published by Coxe in the *Philosophical Transactions* in November 1674, containing detailed descriptions of vitriols, besides some general remarks on the texture of the particles of sulphur.

He developed several of Boyle's chemical ideas and experiments at the Royal Society in the 1660s and 1670s – like Hooke did with pneumatics.

The application of chemistry to the study of plant physiology was central to Nehemiah Grew's research. His papers read at the Royal Society dealt extensively with chemical theory, notably with the question of principles. On 10 December 1674 Grew read a paper on mixture, in which he expounded his theory of matter. Grew stated that the ultimate principles of bodies are atoms, having different sizes and figures. He maintained that different arrangements of atoms produce different mixed bodies and attempted to explain how textures are produced by the different combinations of atoms. Atoms are combined in three different ways: congregation, when they touch in a point (the mixture is not stable); union, when "they touch in a plain" (as in crystals and salts); concentration, "when two, or more atomes touch by reception and intrusion of one into another" (the most compact and fixed bodies).[41] Grew also proposed six causes of the generation of mixed bodies: congruity, weight, compression, solution, digestion and agitation.[42] Grew's adherence to a mechanical theory of matter does not entail the reduction of chemistry to mechanics. Though chemical substances are ultimately composed of insensible particles, the chemical reactions are explained on the grounds of the chemical properties of different substances. No reduction to the mechanical properties of corpuscles is attempted by Grew.

Grew's adherence to the mechanical theory of matter did not bring about his rejection of chemical principles. In *The Anatomy of Plants* Grew did not share van Helmont's and Boyle's view that the principles are generated by fire. He saw salt as the principle from which vegetables are generated – though he recognised that this term comprises different kinds of substances.[43] Grew dealt with the *vexata quaestio* of the pre-existence of alkaline salt in vegetables. His position was that "there is an alkalizate Salt existent in many Vegetables, even in their natural estate; and that it is not made Alkalizate, but only Lixivial by the fire."[44]

[41] N. Grew, 'A Discourse read before the Royal Society Dec 10 1674 Concerning the Nature, Causes, and Power of Mixture', in *Several Lectures Read before the Royal Society* (London, 1682), pp. 223-9.

[42] Ibid., pp. 229-31.

[43] N. Grew, *The Anatomy of Plants* (London, 1682).

[44] N. Grew, *Experiments in Consort of the Luctation Arising from the Affusion of several Menstruums upon all sorts of Bodies* (London, 1678), p. 9. This paper was read at the Royal Society on 13 April and 1 June 1676.

CHAPTER 6

CORPUSCULAR CHEMISTRY IN THE LAST DECADES OF THE SEVENTEENTH CENTURY

INTRODUCTION

In the present chapter I set out to investigate the chemists' theories of matter and the *fortuna* of Boyle's chemical ideas in late seventeenth-century Europe. Though Gassendi's and – to a lesser extent – Descartes's theories of matter influenced late seventeenth-century chemistry, the impact of Boyle's corpuscular philosophy on continental chemistry was far from insignificant. Given the unsystematic character of his works, their impact in Europe is difficult to assess. Nonetheless, if we confine our investigation to his chemical ideas, the influence of Boyle (which I have already assessed for England), was by no means negligible. Reception of his ideas was diverse and not confined to the practical aspects of his chemistry. As we shall see, in the last decades of the seventeenth century, mainly in France, a number of chemists adopted corpuscular ideas, and only few of them reduced chemical properties to the mechanical principles. Moreover, as a result of Boyle's criticism of the chemical principles, several chemists adopted the so-called principles as 'working tools', and did not consider them as the ultimate constituents of all bodies. This is apparent mainly in the numerous textbooks produced in the last decades of the century. The present chapter takes into account the European chemists' theories within their national and intellectual contexts. There are in fact substantial differences between the scientific communities in various parts of Europe, which affected the development of chemical ideas towards the end of the seventeenth century. In France chemistry was mainly developed within the *Académie des Sciences*, but in Germany and in the Netherlands it became part of the university curriculum. In seventeenth-century Italy chemistry was still marginal in the scientific community, except for the Neapolitan *Accademia degli Investiganti*, and then mainly among physicians. If we move to the different intellectual traditions, we may see that both in Germany and in Italy Helmontian themes survived until the end of the century. Boyle's influence was often combined with van Helmont's. Corpuscular and iatrochemical views were very closely linked. Descartes's influence on French and Dutch chemistry was relevant. A number of chemists embraced part of Cartesian natural philosophy – notably the definition of matter and the ether – but, with the exception of Hartsoeker (who was in fact a physicist), Descartes's strong reductionist programme was not adopted. Chemistry retained its own role as an independent discipline even among the members of the *Académie des Sciences*. A qualitative version of corpuscular philosophy became acceptable among French chemists, who were aware of the importance of Boyle's argument against the chemical principles. Hence, we find that chemists like Lemery and Homberg adopted the chemical principles in their investigations, but did not see them as the ultimate constituents of mixed bodies.

THE CARTESIAN HERITAGE IN FRANCE

We have already taken into account Descartes's view of salt as stated in the *Météores,* where he described saline corpuscles along purely mechanical lines.[1] Though there is evidence that between 1620 and 1630 Descartes was interested in the medical uses of chemistry,[2] after the 1630s he unambiguously dismissed chemistry in both his correspondence and his works. His main criticism was the obscurity and unintelligibility of the chemists' (and alchemists') terminology.[3] No less important was Descartes's argument that the chemists conceived as principles what were simply different forms of particles – of one single homogeneous matter. This mistake was for Descartes the outcome of false imagination ('fausse imagination'). As he put it in a letter to Mersenne of 30 July 1640:

> Vous nommez le Sel, l'Huile et le Souffre, pour les Principes des Chimistes; où vous mettez l'Huile au lieu du Mercure, car ils prennent l'Huile et le Souffre pour mesme chose, comme aussi l'Eau et le Mercure. Or ces principes ne sont rien qu'une fausse imagination, fondée sur ce qu'en leurs distillations ils tirent des eaux, qui sont toutes les parties plus glissantes et pliantes des cors dont ils les tirent, et ils les rapportent au Mercure. Ils en tirent aussi des huiles, dont les

[1] See above, chapter 2, pp. 54-5.

[2] In a letter to Mersenne of 15 April 1630 Descartes wrote: "J'estudie maintenant en chymie & en anatomie tout ensemble, et apprens tous les iours quelque chose que ie ne trouve pas dedans les livres." AT, i, p. 379. Cf. J-F. Maillard, 'Descartes et l'alchimie: une tentation conjurée?' in F. Greiner (ed.), *Aspects de la tradition alchimique au XVII*[e] *siècle* (Paris-Milan, 1998), pp. 95-109.

[3] AT, v, p. 237 (letter to Charles Cavendish of 31 March 1649) and AT, vi, p. 9 (*Discours de la Méthode*).

> parties sont en forme de branches, assez deliées et pliantes pour pouvoir estre separées, et ils les rapportent au Souffre; et ils rapportent au Sel les parties plus deliées de ce qui reste, qui se peuvent méler et comme incorporer avec l'eau; puis enfin les parties plus grossieres qui demeurent, sont leur Caput mortuum ou Terra damnata, qu'ils ne content que comme une chose inutile. Au reste, je ne conçoy point ces parties indivisibles ny autrement differentes entr'elles, que par la diversité de leur figure.[4]

In the *Principia* (iv, 63) Descartes suggests an analogy between what he styles acid juices, oils and mercuries and the three chemical principles.[5] It is however apparent that they are not given distinct chemical properties. Like all natural agents, they have mechanical properties only (magnitude, figure and motion) and act in a purely mechanical way. Both in *Le Monde* and in the *Principia* Descartes refers to three elements (fire, air and earth), corresponding to three different kinds of matter. In fact they differ only in the shape and size of the particles. All that is left of the elements are just their names.

In the fourth part of *Principia*, where Descartes deals with chemical and geological phenomena, he adopts a strictly reductionist attitude, explaining all in mechanical terms. The particles of water are of two kinds, one soft and flexible (sweet water), the other hard and rigid (salt water), both being long and conjoined. Salt is composed of the bigger particles of sea water, being equally pointed at both ends.[6] Salt liquefies in water because the slippery and flexible water corpuscles wrap themselves around the sharp points of the salt and carry these away with them. Vitriol, alum and minerals, which are sharp and corrosive, are composed of particles having the shapes of little blades. They have been generated like blades from an iron bar being beaten out by collision with other particles.[7] The particles of oils, sulphur, bitumen and other fatty oily minerals are softer and split into thin and flexible

[4] AT, iii, pp. 130-131. On Descartes's view of imagination, see J-R. Armogathe, 'L'imagination de Mersenne à Pascal', in M. Bianchi and M. Fattori (eds.), *Phantasia-Imaginatio* (Rome, 1988), pp. 259-72.

[5] "Tria hic habemus, quae pro tribus vulgatis Chymicorum principiis, sale sulphure ac mercurio sumi possunt: sumendo scilicet succum acrem pro sale, mollissimos ramulos oleaginae materiae pro sulphure, ipsumque argentum vivum pro illorum mercurio." (AT, viii, p. 241).

[6] *Principia*, iv, p. 66, AT, viii, p. 244.

[7] *Principia*, iv, p. 61, AT, viii, p. 241.

branches.[8] The generation of different kinds of salts is a purely mechanical process: particles of sea salt pass through the pores of earth and have their form changed, so they are transformed into saltpetre, salt ammoniac or other sorts of salt. Nitre is composed of long and rigid particles, pointed more at the one end than to the other.[9]

As the main feature of Descartes's philosophy was to give an account of the generation of natural phenomena as well as of natural bodies from the various mechanical arrangements of the homogenous and inert matter, one can hardly expect to find a description of the specific properties of bodies. If on occasions one does in fact find the description of a limited range of physical phenomena, chemical ones are always reduced to the mechanical affections of corpuscles. It is therefore safe to conclude that in Descartes's natural philosophy there is no place for chemistry as an independent discipline.[10]

This was not the case with the French and Dutch chemists, who often adopted a moderate version of Descartes's mechanism, which gave chemistry an independent status. This was due to the combined influence of both Gassendi's and Boyle's corpuscular doctrines – both of them being of relevance among the members of the *Académie des Sciences*.

THE THEORY OF PRINCIPLES IN SEVENTEENTH-CENTURY FRENCH CHEMICAL TEXTBOOKS

The rise of the chemical textbook in the seventeenth century was primarily a French phenomenon. Jean Beguin's *Tyrocinium* (heavily indebted to Libavius) set out a tradition which spread all over Europe. French textbooks were translated into other languages and provided a model for chemical courses produced in other countries.[11]

[8] *Principia*, iv, p. 62, AT, viii, p. 242.

[9] *Principia*, iv, pp. 69 and 110, AT, viii, pp. 245 and 264.

[10] In his physiological work Descartes adopts the notion of fermentation, though just as an analogy: he claims that vital heat is produced by a process which is analogous to fermentation. See *Traité de l'Homme*, AT, xi, p. 123 and the letter to Plemp of 15 February 1638, AT, v, pp. 530-1.

[11] A quantitative analysis of the contents of some French chemical textbooks may be found in M. Bougard, *La Chimie de Nicolas Lemery* (Thurnhout, 1999), pp. 418-28. On chemical textbooks, see also B. Joly, 'De l'alchimie à la chimie: le dévelopment des "cours de chymie" au XVIIe siècle en France', in Greiner (ed.), *Aspects* (n. 2), pp. 85-94 and L. Principe, *The Aspiring Adept. Robert Boyle and his Alchemical Quest* (Princeton, 1998), pp. 34-5 and 58-61.

Most seventeenth-century chemical textbooks were aimed at giving practical instructions to pharmacists and physicians and paid little attention to the theoretical part of chemistry. This was certainly the case of Beguin's *Tyrocinium* and of its numerous (revised) editions in different languages.[12] In *The Sceptical Chymist*, as I have argued elsewhere, Boyle's attack was directed against the authors of chemical textbooks like the *Tyrocinium* – not against the alchemists. For Boyle the textbooks were 'unphilosophical', that is, a collection of recipes introduced by the traditional doctrine of the chemical principles.[13] Yet, it would be a mistake to suppose that all chemical textbooks were purely practical in their contents. Some chemical courses published in the second half of the seventeenth century did not confine themselves to the traditional exposition of chemical principles. They also include substantial theoretical sections. Corpuscular theories (sometimes echoing Descartes) are taken into account and Boyle's theories are favourably reported. Evidently, the different emphasis on the theoretical aspects of chemistry depended both on the putative readership of the course and on the intellectual standing of the author. Here I investigate only some of the most popular French chemical courses, focusing on the chemical theories therein contained.

As we have already seen, de Clave's *Course* goes beyond the traditional limitations of chemical textbooks, since it deals thoroughly with the theory of matter.[14] Annibal Barlet's course, first published in 1653, contains a large cosmological section, a digression on angels, and also refers to atoms. Barlet (who rejects the existence of vacuum and the plurality of the worlds) describes the action of fire as a decomposition of bodies to their atoms.[15] In the case of sublimation, according to Barlet, the most subtle atoms depart from the body due to the action of heat.[16] He believes that the principles of natural bodies are two, the universal spirit and salt. The spirit is described as "une substance subtile et rare", while salt is solid and compact. From these

[12] On J. Beguin's *Tyrocinium* (1610) see T.S. Patterson, 'Jean Beguin and his Tyrocinium Chymicum', *Annals of Science* 2 (1937), 243-298 and O. Hannaway and A. Kent, 'Some new considerations on Beguin and Libavius', *Annals of Science* 16 (1960), 241-50.

[13] A. Clericuzio, 'Carneade and the Chemists', in *Robert Boyle Reconsidered*, pp. 82-3. See also Principe, *Aspiring Adept* (n. 11), pp. 58-62.

[14] See above, chapter two, pp. 42-7.

[15] A. Barlet, *Le Vray et methodique cours de la physique resolutive, vulgairment dite Chymie* (Paris 1657^2), pp. 66-7.

[16] Ibid., p. 117

two substances originate the chemical principles: fire, sal ammoniac, water, mercury, sulphur, earth and salt (common salt).[17]

As may be seen from Nicaise Lefebvre's *Traicté de la Chymie* (1660), the status of chemistry in 1660 was already an important issue in French science. Lefebvre opens his *Traicté* with the question of whether chemistry is art or science. His answer is that chemistry is divided into three parts, each having a different status. The first is philosophical chemistry, which is a science dealing with the principles of nature; the second is iatrochemistry, which is practical, but its principles derive from the philosophical part; finally, there is pharmaceutical chemistry, entirely operative, but subordinated to iatrochemistry.[18] One task of chemistry, according to Lefebvre, is to establish the number of principles and to extract them. Its other task is the one upon which Paracelsus and Severinus had insisted, namely, spiritualising bodies and corporifying spirits: "la chymie ne montre pas seulement comment le corps peut estre spiritualisé: mais elle montre aussi comment l'esprit se corporifie."[19] The first part of the book deals extensively with the universal spirit, a simple and homogeneous substance, which is the source of life in the universe. Along with the Paracelsians, Lefebvre claims that the spirit is specificated into individual substances by ferments, according to the different matrices which receive it:

> Or comme cet esprit est universel, aussi ne peut il estre specifié que par les moyen des ferments particuliers qui impriment en luy le caractere & l'idée des mixtes, pour estre faits tels ou tels êtres determinez, selon la diversité des matrices qui reçoivent cet esprit pour le corporifier. Ainsi dans une matrice vitriolique, il devient vitriol, dans une matrice arsenicale, il devient Arsenic, la matrice vegetable le fait estre plante.[20]

Salt, sulphur and mercury are the active principles, water and earth are not real principles but elements, as they can be transformed one into the other. For Lefebvre, they all derive from the spirit of the world. The extraction of the chemical principles is a major task of the chemist. The

[17] Ibid., pp. 42 and 52-3.

[18] N. Lefebvre, *Traicté de la Chymie*, 2 vols. (Paris, 1660), i, pp. 6-11. On Lefebvre see Partington, iii, pp. 17-24.

[19] Lefebvre, *Traicté* (n. 18), p. 15.

[20] Ibid., p. 19. Cf. N.E. Emerton, *The Scientific Reinterpretation of Form* (Ithaca and London, 1984), pp. 186-7.

three principles which are used in medicine must be freed from the impure parts they contain.[21]

In the *Traité de la Chymie*, published in 1663 by Christophle Glaser (professor of chemistry at the Jardin du Roy after Lefebvre) the theoretical part is much shorter than the one in Lefebvre's *Traicté*.[22] Like Lefebvre, Glaser gives prominence to spirit as the most active substance. The cosmological speculations about spirit which we have seen in Lefebvre, disappear from Glaser's book. The definition of spirit is more operational: it is one of the three principles, the first substance to be extracted and the origin of motion in natural bodies.[23] It easily volatilises, and, as a consequence, those substances abounding with spirit are not durable. Sulphur is the second principle and the link between spirit and salt ("il fait la liaison des autres principes, lesquels sans luy ne se pourroient entretenir").[24] It is less active than spirit and is the source of colours, odours, malleability, ductility, etc. Salt is fixed and gives solidity to bodies, besides preserving them from corruption.[25]

Pierre Thibaut's *Cours de Chymie* (1667) and Sébastien Matte-La Faveur's *Pratique de Chymie* (1671) are mainly practical in orientation.[26] Thibaut purposely avoided dealing with chemical theories, which he deemed as obscure and useless. In Thibaut's *Cours* we do not even find the traditional description of the Paracelsian principles and of their properties. Matte-La Faveur defines chemistry as an art and has little to say about the principles. Nevertheless, his *Pratique* has more chemical theory than Thibaut's *Cours* and some of his statements on the properties of chemical principles show that a substantial change in the view of the principles is occurring. The five principles have lost their simplicity. There are different kinds of salts, sulphurs and mercuries, each having different qualities.

[21] Ibid., pp. 21-6.

[22] On Christophle Glaser, see R.G. Neville, 'Christophle Glaser and the *Traité de la Chymie*, 1663' *Chymia* 10 (1965), 25-52 and Partington, iii, pp. 24-6. Moïse Charas claimed that he (not Glaser) was the author of the *Traité de la Chymie*, but there is no evidence to confirm Charas's claim. See M. Bougard, *La Chimie* (n. 11), pp. 24-6.

[23] Glaser, *Traité de la Chymie* (Paris, 1668; 1st edn: 1663), p. 7.

[24] Ibid., p. 8.

[25] Ibid., p. 9.

[26] Very little is known of Pierre Thibaut. Evidently he originated from Lorraine. From a laudatory sonnet published in his *Cours* we know that he was 'Distillateur ordinaire du Roy'.

Matte-La Faveur adopts the traditional Aristotelian distinction between perfect and imperfect mixts, which he reinterprets in vague corpuscular terms. Imperfect mixts, he says, are those which can easily be decomposed by heat; perfect ones are those which resist the action of heat, because their parts have a closer union.[27] The chemical principles, he asserts, are not the same in the three kingdoms of nature. Volatility, he states, is a relative quality. Though mercury is the most volatile among the principles, salts too are volatile, though in different degrees. The same principle can be more or less volatile, according to the natural kingdom from which it originates. The principles of minerals are the most fixed, those extracted from vegetables are the most volatile.[28]

Little or no reference to the corpuscular theory of matter is to be found in the chemical courses we have examined.

In any comparison of these chemical textbooks, Lemery's is paramount both for its very substantial theoretical section and for its adoption of a full-scale corpuscular theory of matter. Since Nicolas Lemery's chemistry has been thoroughly examined by Bougard, I here confine myself to an investigation of only two aspects of Lemery's work, his use of the corpuscular theory in chemistry and its relationship to Boyle's chemistry.

Nicolas Lemery's *Cours de Chymie*, first published in 1675, went through a number of editions which contain substantial differences from the

[27] Sébastien Matte-La Faveur, *Pratique de Chymie* (Montpellier, 1671), pp. 4-5. Matte-La Faveur was 'Distillateur and Demonstrateur ordinaire de la Chymie' at the Medical faculty of Montpellier. On Matte-La Faveur see M. Bougard, *Autour de Sébastien Matte-La Faveur. Eclaircissement biographiques sur une famille de démonstrateurs de la chimie à Montpelliers (XVIIe et XVIIIe siècles* (n. p., 1989), and id., *La Chimie* (n. 11), pp. 126-7.

[28] "Dans un sens relatif un principe est dit volatil ou fixe par rapport aux autres parties du même mixte, ou l'on compare des principes de même nature, par exemple deux differents sels d'un même mixte; ou des principes de differente nature, par exemple le sel & le soufre d'un même mixte. Dans le premier sens on dit qu'un sel est fixe & que l'autre est volatil. Dans le second sens on dit que le mercure est le principe le plus volatil du mixte. Si vous comparez un principe avec celuy d'un autre mixte, vous le pourrez appeler volatil ou fixe selon la fixité, ou la fixité du principe avec qui vous le comparerez. Ains le sel essentiel des plantes, ou le nitre comparé avec le sel armoniac, ou avec le sel volatil de quelque animal, est dit fixe; quoyque le comparant avec les Alkalis il soit volatil. Avant que de passer aux differences particulieres de chaque principe, remarquez que les mineraux ont leurs principes plus fixes que le vegataux & derechef les vegetaux plus que les animaux." Matte-La Faveur, *Pratique de Chymie* (n. 27), pp. 13-14.

first edition.[29] At the outset Lemery, along with Descartes and Boyle, complains of the obscurity of the current chemical terminology, which in his view has contributed towards delaying the progress of this discipline.[30] Lemery's arguments are evidently intended to legitimise chemistry among natural philosophers by vindicating its role in the face of persistent opposition and scepticism. The theory of five principles as proposed in the first edition is traditional. Lemery distinguishes three active principles and two passive ones, but he distances himself from Lefebvre's theory of the spirit of the world as the active vital principle since this notion is more metaphysical than physical.[31] Lemery's spirit (or mercury) is described according to its chemical properties: it is volatile, it is the first substance to

[29] Bougard, *La Chimie* (n. 11), pp. 439-55, publishes the text of Lemery's 'Remarques sur les Principes' from the 1687 edition of the *Cours*, which is to some degree different from the first edition (1675). The 1687 text is not changed in subsequent editions of the work. For a bibliography of Lemery's works and a comparative analysis of the different editions of the *Cours*, see Bougard, *La Chimie* (n. 11), pp. 393-435. On Lemery see also J-C. Guédon, 'Protestantisme et Chimie: Le milieu intellectuel de Nicolas Lémery', *Isis* 65 (1974), 212-228 and J.C. Powers, 'Ars sine Arte: Nicholas Lemery and the End of Alchemy in Eighteenth-Century France', *Ambix* 45 (1998), 163-89.

[30] "La pluspart des Autheurs qui ont parlé de la Chymie, en ont écrit avec tant d'obscurité, qu'ils semblent avoir fait leur possible pour n'estre pas entendus. Et l'on peut dire qu'ils ont trop bien réüssi, puisque cette Science a esté presque cachée pendant plusieurs siecles, & n'a esté connuë que de tres-peu de personnes. C'est en partie ce qui a empesché un plus grand progrés que l'on eust pû faire dans la Philosophie." N. Lemery, *Cours de Chymie* (Paris, 1675) (hereafter as *Cours*, followed by the date of edition), 'Preface' sig. aiij^r. Descartes's criticism of the chemists' language appears in his letter to Newcastle of 23 November 1646: "Je souscris en tout au jugement que Vostre Excellence fait des Chimistes, & croy qu'ils ne font que dire des mots hors de l'usage commun, pour faire semblant de sçavoir ce qu'ils ignorent." AT, iv, pp. 569-70. Boyle's less destructive criticism is contained in the *Sceptical Chymist*, see above, pp. 114-5. On the language of chemistry see M. Crosland, *Historical Studies in the Language of Chemistry* (New York, 1978²) and M. Beretta, *The Enlightenment of Matter. The Definition of Chemistry from Agricola to Lavoisier* (Canton, Mass., 1993).

[31] "Le premier Principe qu'on peut admettre pour la composition des Mixtes est un esprit universel, qui étant répandu par tout, produit diverses choses selon les diverses Matrices ou Pores de la Terre dans lesquels il se trouve embarassé: mais comme ce Principe est un peu Metaphysique, & qu'il ne tombe point sous les sens, il est bon d'en établir de sensibles." Lemery, *Cours* (1675), p. 3 (this passage is not changed in subsequent editions).

be extracted by fire analysis and the principle of activity in bodies. It is found in great quantity in plants and animals, but is rather scarce in minerals.[32] The description of the other principles contains nothing innovative. What is original is Lemery's definition of the principles, which appears in a section entitled 'Remarques sur les Principes', which he introduced in the 1683 edition of his *Cours*, and did not subsequently change.[33] As I have argued elsewhere, Lemery's changing view of principles is evidently a reply to Boyle's criticism of the chemical doctrine of principles as contained in *Producibleness of Chemical Principles* (appended to the second edition of the *Sceptical Chymist*, 1680).[34] Lemery does not accept Boyle's view that the five principles are produced, not extracted, by fire, though he confirms that there are substances (like gold and silver) from which it is impossible to extract the principles. In addition, he asserts that volatile salts obtained by distilling plants do not pre-exist in them. They have been produced by fire.[35] Lemery also recognises that the principles can

[32] "L'Esprit qu'on appelle Mercure, est le premier des actifs, qui nous paroist lorsque nous faisons l'Anatomie d'un Mixte: c'est une substance subtile, penetrante, legere, qui est plus en agitation qu'aucun des autres Principes: c'est luy qui fait croistre les Mixtes en plus ou en moins de temps; selon qu'il s'y rencontre en petite ou en grande quantité: mais aussi pour son grand mouvement, il arrive que les corps où il abonde, sont plus subject à la corruption; c'est ce qu'on remarque aux Animaux, & aux Vegetaux. Au contraire, la plûpart des Mineraux où il est en petite quantité, semblent incorruptibles." Ibid., p. 3.

[33] Bougard, *La Chimie* (n. 11), pp. 439-55 publishes the text of Lemery's 'Remarques sur les Principes' as contained in the 6th edition of the *Cours* (1687), which is rather different from the first edition (1675). The 1687 text is not changed in subsequent editions of the work.

[34] Clericuzio, 'Carneades and the Chemists', *Robert Boyle Reconsidered*, pp. 84-5.

[35] "Quelques philosophes modernes veulent persuader qu'il est incertain que les substances qu'on retire des mixtes, & que nous avons appellées Principes de Chymie, resident effectivement & naturellement dans le Mixte: ils disent que le feu en rarefiant la matiere dans les distillations, est capable de luy donner ensuite un arrangement tout different de celuy qu'elle avoit auparavant, & de former le Sel, l'Huile & les autres choses qu'on en titre. Ce doute paroist d'abord assez bien fondé; parce qu'il est certain, comme nous le dirons dans la suite, que le feu donne beaucoup d'impression aux preparations, & que bien souvent il deguise tellement les substances, qu'elles ne sont presque plus reconnoissables de ce qu'elles étoint auparavant, mais, il est facile de faire voir que quoy que le feu déguise les substances, il ne forme pas neanmoins les Principes; car nous les voyons et sentons dans plusieurs Mixtes avant qu'ils ayent passé par le feu.", Lemery, *Cours* (1683),

be so closely linked the one to the other that they cannot be recovered in their pristine form.[36] Nonetheless, Lemery does not reject the doctrine of principles, though he adopts it in a weak form which reveals his acceptance of some of Boyle's arguments against the chemists' principles. Lemery asserts that he employs the term principle as a 'working tool', not in a strict sense, since the substances commonly called principles could be further divided. As he puts it, they are principles for us, not in nature:

> Le nom de Principe en Chymie, ne doit pas estre pris dans une signification tout à fait exacte; car les substances qu'on appelle ainsi, ne sont Principes qu'à nostre égard, & qu'entant que nous ne pouvons point aller plus avant dans la division des corps, mais on comprend bien que ces Principes sont encore divisibles en une infinité de parties qui pourroient, à plus juste titre, estre appellées Principes. On n'entend donc par Principes de Chimie que des substances separées & divisées autant que nos foibles efforts en sont capables.[37]

On occasions, Lemery explains chemical reactions in mechanical terms. The reference to the shape of particles shows Lemery's adherence to Descartes's theory of matter, rather than to Boyle's. In a typical Cartesian way, Lemery maintains that the properties of acid substances are produced by their pointed particles, as is confirmed by their forms when they crystallise. Alkalis owe their properties to the fact that their particles are brittle and have pores with a shape which fits the points of acids' particles.

pp. 6-7. "On trouve aisément les cinq Principes dans les Animaux & dans les Vegetaux, mais on ne les rencontre pas avec la mesme facilité dans les Mineraux: il y en a mesme quelques-uns, comme l'Or et l'Argent, desquels on ne peut pas en tirer deux, ny faire aucune separation, quoy que nous disent ceux qui recherchent avec tant de soin, les Sels, les Soulfres & les Mercures de ces Metaux." Ibid., p. 9. See also ibid., p. 21, on the volatile salts of plants.

[36] "Je veux croire que tous les Principes entrent dans la composition de ces Mixtes, Mais il n'y a pas de consequence que ces Principes soient demeurez en leur premier état, & qu'on les en puisse retirer; car il se peut faire que ces substances qu'on appelle Principes, se soient tellement embarassées les unes dans les autres, qu'on ne les puisse pas separer qu'en brisant leurs Figures. Si vous meslez par exemple un Esprit acid avec le sel de tartre ou quelqu'autre alkali, le pointes de l'acide s'embarrassent de maniere dans les Pores du Sel; que si par la distillation vous voulez separer l'esprit acide comme il estoit auparavant, vous n'y parviendrez jamais: il aura perdu presque toute sa force, parce que ses pointes s'estant brisées dans l'effort qu'elles auront fait, n'auront peû conserver la figure aussi penetrante qu'elles avoient." Ibid., p. 9.

[37] Ibid., p. 6.

The different degree of corrosiveness of acids depends on the sharpness of their points.[38] Lemery adopts mechanical views in the section on mercury:

> Les parties du mercure estant supposées rondes, on peurra expliquer comment ce métal demeure fluide et pourquoy il est si facilement volatilizé par le feu quoy qu'il soit fort pesant, car la figure ronde n'estant nullement propre a la liaison des parties, les petits corps qui composent le vif-argent ne peuvent estre unis entr'eux & par consequence ils doivent rouler les uns sur les autres.[39]

Lemery however does not push the mechanical explanations too far. He states that "ce n'est qu'à raison de leurs figures qu'ils [les principes] peuvent être dits Sels, Soulfres et Esprits", but he fails to give an exact description of their figures. He says only that alkaline salts have pores and that the particles of oils are flexible.[40] It is notable that in the early editions of the *Cours* Lemery speaks of fire corpuscles (a qualitative version of corpuscularianism) and changes his view of fire only in the 1687 edition, introducing a Cartesian definition of fire: "le feu ordinaire n'est qu'un mouvement tres-violent de petits corps autour de leur centre."[41] Though Lemery adopts relevant aspects of Descartes's mechanism, it is apparent that his chemistry is not wholly based on the Cartesian theory of matter.

Two more aspects of Lemery's *Cours* bear witness to his adoption of Boyle's chemical views. The first is the attempt to propose a new classification of chemical substances; the second, related to the former, is the quest for a reform of chemical terminology. Lemery's classification of chemical substances is not entirely new. As we have seen, other chemical courses tried to differentiate salts and to establish differences between the *tria prima* extracted from plants and those obtained from animals. Lemery classifies three different kinds of salts, namely fixed salts (mainly those obtained from plants by means of calcination), volatile salts (from animals), and essential salts (obtained from vegetable juices).[42] According to Lemery, the three spirits ("esprit des animaux, esprit ardent des vegetaux, esprit

[38] Ibid., pp. 25-6.

[39] Ibid., p. 169. Equally mechanical is Lemery's description of cinnabar (mercuric sulphide): "La cause de ce déguisement du Mercure en Cinabre vient de ce que la partie du soulfre la plus acide penetre le Mercure, & embarasse tellement ses parties, qu'elle arreste l'agitation en laquelle elles estoient." Ibid., p. 172.

[40] Ibid., p. 9.

[41] Lemery, *Cours* (1687), p. 582. See also Bougard, *La Chimie* (n. 11), pp. 166-7.

[42] Lemery, *Cours* (1687), p. 4.

acide") are in fact salts or oils.[43] Though Lemery complains, like Boyle, that the term spirit has generated much confusion in chemistry, he maintains that it is useful to keep the traditional term of spirit, with the caveat that one has to know that it denotes not a specific chemical substance, but a variety of different bodies.[44]

ACID AND ALKALI IN FRANÇE: ANDRÉ AND BERTRAND

The acid/alkali doctrine, which stemmed from van Helmont, was articulated by Sylvius (1663) and Tachenius (1666), and subsequently by François André (1672). As we have seen, Boyle attacked it in 1675.[45] This doctrine was later vindicated by André, who in 1677 issued a second edition of his *Entretiens sur l'Acide et sur l'Alkali*, to which he added a reply to Boyle. The acid/alkali theory received a new attack from Bertrand, whose arguments show the impact of Boyle's objections.

The second edition of André's *Entretiens* (1677) contains an answer to Boyle's main objection, namely that the definition of these principles is circular, as they cannot be identified separately, but of necessity one by means of the other. André tries to attribute distinct properties to each of them. He maintains that the acid salt is often found in the liquid state, whereas the alkaline one is in a solid state.[46] André's two principles do not have the same status. The acid salt is the active one on which the origin of bodies depends: it is described as "l'auteur de la construction de tous les

[43] "Il y a de trois sortes de Liqueurs, qu'on qualifie du nom d'Esprit dans la Chymie, l'Esprit des Animaux, l'Esprit ardent des Vegetaux, & l'Esprit Acide. Le premier, comme l'Esprit de Corne de Cerf, n'est qu'un Sel Volatile Resout par un peu de Phlegme. Le seconde, comme l'Esprit de Vin, l'Esprit de Genievre, l'Esprit de Romarin, est une Huile exaltée, comme nous dirons en parlant des Vins. Et le dernier, comme l'Esprit de Vinaigre, l'Esprit de Tartre, l'Esprit de Vitriol, est un Sel essentiel Acide Resout en fusion par le feu, comme nous prouverons en parlant du Vinaigre et de la distillation du Tartre. On appelle cette derniere sorte d'Esprit, Sal fluor, parce qu'en effet, ce n'est qu'un Sel fluide. Ce trois sortes de liqueurs comprenant tout ce qu'on appelle Esprit." Ibid., pp. 9-10.

[44] Ibid., p. 11.

[45] François de le Boë (Sylvius), *Disputationum Medicarum Decas* (Amsterdam, 1663) and O. Tachenius, *Hippocrates Chimicus* (Venice, 1666). On Tachenius see Partington, ii, pp. 291-7. On Sylvius, see DSB and E.A. Underwood, 'Franciscus Sylvius and his Iatrochemical School', *Endeavour* 31 (1972), 73-6.

[46] F. André, *Entretiens sur l'Acide et sur l'Alkali. Où sont examinées les objections de Mr Boyle contre ces principes* (Paris, 1677, first edn: 1672).

corps." André's definition of the acid salt is based on the mechanical properties of its particles:

> Le sel acide se connoist facilement au goust, à l'odeur, & par la fermentation qu'il fait avec les alkali, comme l'esprit de souffre. Ce sel est composé de petites parties pointües, lesquelles s'insinuent dans les pores des corpes qu'elles rencontrent & en font ou la desunion des parties ou la coagulation.[47]

The alkali principle, which is subordinate to the acid one, is porous and its particles have different shapes. Answering Boyle's objections, André defends the acid/alkali doctrine by stating that it is based on notions which are simple and universal. Boyle had in fact criticised this theory because it was too comprehensive, since there are substances which are neither acid nor alkaline. André denies that its comprehensiveness is a sign of weakness and rejects Boyle's experimental arguments for the existence of neutral substances. In addition, André claims, the acid/alkali theory is not incompatible with the principles of mechanical philosophy. In fact he refers the properties of both these principles to the form, size and motion of particles.[48]

In 1683 the acid/alkali theory found a new advocate in Bertrand, physician from Marseilles.[49] Echoing Boyle, Bertrand objects to André that one cannot build a system of natural philosophy on the acid/alkali theory, which, as it is commonly expounded, is "trope vague et obscure".[50] Not all chemical reactions, Bertrand believes, can be explained by acid and alkali. A number of facts disprove the postulated universality of this doctrine. Firstly metals are not composed of acids and alkalis; secondly the generation of plants cannot be explained by these two principles; thirdly there are several substances which clearly do not contain them; finally the properties of opening other bodies' texture is not a specific property of acids, as other substances can produce the same effect.[51] Bertrand concludes his objections maintaining that, even if they were contained in all bodies,

[47] Ibid., p. 15.

[48] Ibid., pp. 103-5, 131-5.

[49] Little is known about Bertrand, not even his Christian name. He was a member of the Marseilles College of Physicians.

[50] Bertrand, *Reflexions Nouvelles sur l'Acide et sur l'Alkali: où apres avoir demontré que ces deux sels ne peuvent pas être les principes des Mixtes, on fait voir le veritable usage qu'on en peut faire dans la Physique & dans la Medecine* (Lyons, 1683), pp. 2-3.

[51] Ibid., pp. 16; 23, 25; 42; 45-6; 59-72; 76.

this would not imply that they must be principles. They may be "un assemblage d'autres choses plus simples."[52] Bertrand, who explicitly refers to Lemery, accorded acid a special status. He states that there is in nature a universal acid which produces metals and vegetables.[53] Acids are responsible for most human diseases.[54] He explains the properties of acids and of alkali doctrine on the grounds of the mechanical properties of the corpuscles. Acids are liquid substances with small and pointed particles. Therefore, acid substances open other bodies and produce effervescence. The different forms of their spikes account for the variety of acids and for their different chemical properties. Acid easily penetrates alkali because the latter has a porous texture.[55] In conclusion, Bertrand denies that alkali can be regarded as principles. He deems them a re-arrangement of particles due to fire. Bertrand recognises that the mechanical properties of bodies cannot be disclosed by means of senses, but can be assumed only by means of imagination and reasoning.[56] Though he put special emphasis on the mechanical properties of corpuscles, Bertrand, in a way typical of Boyle, argues that in natural philosophy, as well as in chemistry and medicine, one has to refer to intermediate causes, closer to the empirical evidence.[57]

THE *ACADÉMIE DES SCIENCES*

Boyle's chemical work was thoroughly examined by the Paris *Académie des Sciences*. Several works of his were discussed in detail by members of the Academy. It is apparent that dissent from Boyle's chemical views generally prevailed, as the Paris *académiciens* did not reject the chemical principles. Nevertheless, Boyle's objections to the chemical theory of matter were by no means ineffective among Parisian scientists, for some members of the *Académie des Sciences* produced a revised version of the chemical principles. Their view was that the five principles were the constituents of bodies, though they were not simple substances.

Duhamel, who was the first Secretary of the *Académie*, dealt extensively with the elements and principles in *De consensu veteris et novae philosophiae* (1663) which went through several editions. Duhamel

[52] Ibid., p. 74.
[53] Ibid., p. 36-8.
[54] Ibid.
[55] Ibid., pp. 3-4; 6.
[56] Ibid., pp. 163-4.
[57] Ibid., pp. 164-8.

expanded his views in *Philosophia vetus et nova*, which was published as a textbook. Here Duhamel expounded a syncretistic theory of matter, trying to reconcile the Aristotelian elements with the corpuscular theory of matter. However, he did not reject Aristotelian forms, which he believed to originate from the matter, where they are contained *in potentia*.[58] The chemical principles cannot account for all the qualities of bodies and are not the ultimate ingredients of bodies, but they derive from the elements. In addition, they are not in the mixed body but are produced by fire. Echoing Boyle, Duhamel stressed that the chemists have given divergent and somewhat confused descriptions of their principles.[59]

The Paris Academy paid special interest to issues related to the theory of matter. Part of the academicians' discussions on this subject stemmed from their examination of Boyle's works. At the Academy's request, Samuel Cottereau Duclos took over the task of producing a detailed analysis of a number of Boyle's chemical works.[60] Part of the Academy's chemical

[58] See Duhamel, *De Consensu veteris et novae philosophiae libri duo*, 3rd edition, revised (Oxford, 1669, first edn: 1663), and id., *Philosophia vetus et nova ad usum scholae accommodata* (London, 1685[4]), pp. 681-742. Jean Baptiste Duhamel (1623-1706) was the first Secretary of the *Académie Royale des Sciences* (1666-97). On Duhamel, see D.J. Sturdy, *Science and Social Status: The Members of the Académie des Sciences, 1666-1750* (Woodbridge, 1995), pp. 82-6.

[59] "Chymicorum principia non prima, sed ad summum principia secundaria dici possunt [...] tanta in iis [chymicis] vocibus salis, sulphuris, & mercurii ambiguitas, tanta est homonymia, ut res his vocibus subjectae vix designari. Principia chymicorum non sunt actu in mixtis, ea forma, numero, statu, in quo separata conspiciuntur: sed ignis magna ex parte ea procreat.", Duhamel, *Philosophia vetus et nova* (n. 58), pp. 743-4.

[60] Samuel Cottereau Duclos (1598-1685), member of the *Académie Royale des Sciences*, King's physician became in 1670 assistant to Bourdelin for the mineral waters tests in France. On his life, see D. Todériciu, 'Sur la vraie biographie de Samuel (Duclos) Cotreau', *Revue d'Histoire des Sciences* 27 (1974), 64-7. See also Sturdy, *Science* (n. 58), pp. 107-9 and Bougard, *La Chimie* (n. 11), pp. 133-6. His *éloge*, written by Condorcet, reads: "Notre chimiste sentit même combien l'application de la Physique corpusculaire à la Chimie était vague et fautive; & il s'éleva hautement contre la Chimie de Boyle, qui était uniquement fondée sur cette physique." *Éloges des Académiciens de l'Académie Royale des Sciences, morts depuis 1666, jusq'en 1699* (Paris, 1773), p. 68. It is noticeable that in his *éloge* of Duclos, Condorcet praises Stahl for making chemistry "une veritable science". Ibid., p. 67. Condorcet's *éloge* is evidently based on Fontenelle's opposition of Duclos's chemical research to Boyle's mechanical views of matter. See *Histoire de l'Académie Royale des Sciences* (Paris, 1733), i, p. 79. Duclos left a substantial

investigations were aimed at determining the principles of mixed bodies, mainly by analysing organic substances. Some of the early academicians (in particular Duclos and Claude Bourdelin) carried out extensive laboratory research to find out the constituents of organic bodies, in particular of plants.[61] Boyle's works were discussed in detail by Duclos. On 31 December 1666 Duclos read a paper on the chemical principles at the Academy. Along with van Helmont, he maintained that water and a spiritual substance are the ultimate principles – which in turn produce the Paracelsian principles.[62] In 1667 Duclos discussed (at Adrien Auzout's suggestion) Boyle's transmutation of gold into silver, as recorded in *The Origin of Forms and Qualities*. When discussing Boyle's transmutation of metals, Duclos maintained that this is not a mechanical process but one produced by a specific ferment.[63] The focus of Duclos's comments was Boyle's *menstruum peracutum*. The academicians also set out to prepare a stronger solvent than Boyle's. This was expected by Duclos to reduce bodies into their ultimate and homogeneous corpuscles.[64] Next, in July 1668, Duclos took into account Boyle's *Certain Physiological Essays* (1661) and objected to Boyle that the nitre he used in his 'redintegration' experiment was no pure substance.[65] In 1677 Duclos read a *Dissertation sur les principes des mixtes*, subsequently published both in the *Mémoires de l'Académie des Sciences* and as a separate book in 1680, where he contends that by fire analysis he extracted water, oil, salt and earth from vegetables. He stressed that the same result is not obtained by the analysis of animals and minerals, which is much more difficult than an analysis of plants. This does not

amount of manuscripts, largely contained in the Procès-verbaux of the *Académie des Sciences*.

[61] See J.G. Stubbs, "Chemistry at the Académie des Sciences" (Unpublished Ph.D. dissertation, University College, London, 1939); F.L. Holmes, 'Analysis by Fire and Solvent Extractions: The Metamorphosis of a Tradition' *Isis* 62 (1971), pp 129-148 and A. Stroup, *A Company of Scientists. Botany, Patronage, and Community at the Seventeenth-Century Parisian Royal Academy of Sciences* (Berkeley, 1990), pp. 89-102.

[62] Académie Royale des Sciences, Procès-verbaux, 22 Dec. 1666, i, fols. 1-22.

[63] Procès-verbaux, i, fols. 113-114 (16 Avril 1667).

[64] Procès-verbaux, i, fols. 108-167, containing observations on the preparation of the *Alkahest*. The reference to homogeneous corpuscles occurs on fols. 140^v-141^r. Duclos's positive view of the *Alkahest* for the analysis of plants was criticised by Denis Dodart (1634-1707), physician and member of the *Académie des Sciences*, see Stroup, *A Company* (n. 61), p. 97, and Sturdy, *Science* (n. 58), pp. 184-9.

[65] Paris, Bibliothèque Nationale, MS fr. 1333, fols. 238-262.

prevent him from generalising about the principles and elements. He seems to have moved from the Helmontian water principle into a revised version of the Aristotelian elements. He states that the substances which compose our world are three, water, earth and air, so these are the ultimate elements of natural bodies. The spiritual principle retains a central role: the elements' movements are directed by light, which in turn is moved by an incorporeal spiritual principle, which is endowed with extension. Duclos claims that some bodies can act one upon the other without physical contact by means of the universal spirit, which surrounds and penetrates them.[66] Nevertheless, Duclos did not dismiss the use of corpuscles, which he evidently did not interpret along mechanical lines. He explained coagulation of milk as a movement of corpuscles which unite to those which have the same nature: "chaque particule se separe de celles qui luy sont heterogènes et se joint aux homogènes."[67]

The chemical analysis of plants was also pursued by Claude Perrault, who along with Boyle aimed at using chemistry to support the corpuscular theory of matter.[68] Perrault's theory of matter is mechanical and his corpuscles differ according to their geometrical forms. Perrault put special emphasis on the activities of the ether, whose thin particles can penetrate all bodies.[69]

Boyle's works were discussed by another member of the *Académie des Sciences*, Edme Mariotte, who also examined the chemistry and physiology of plants.[70] In his *Essay de la Vegetation des Plantes,* based on experiments made at the *Académie des Sciences*,[71] Mariotte considers the chemical principles as compound corpuscles, formed of simpler substances.[72]

[66] Duclos, 'Dissertation', *Histoire et mémoires de l'Académie royale des sciences*, 11 vols. (Paris, 1729-33), iv, pp. 1-40. Duclos's *Dissertation* was also published in Amsterdam in 1680.

[67] Procès-verbaux, vi, fol. 64^{r-v} (27 April 1669).

[68] Procès-verbaux, i, fols. 36-7, 8; 215-6, 222 (1667 and 1678-9). On Claude Perrault (1613-88), physician and Professor of Medicine at the Sorbonne, see Sturdy, *Science* (n. 58), pp. 86-7.

[69] Procès-verbaux, vi, fols. 144-45 (1669). See also C. Perrault, *La Pesanteur des corps*, in C. and P. Perrault, *Oeuvres diverses*, 2 vols. (Leiden, 1721), i, pp. 3-10.

[70] On Mariotte, see DSB; R. Taton (ed.), *Mariotte Savant et Philosophe* (Paris, 1986); Stroup, *A Company* (n. 61), *passim*; and Sturdy, *Science* (n. 58), pp. 110-2.

[71] Mariotte, *Essay de la Vegetation des Plantes* (Paris, 1676), repr. in *Oeuvres de Mr Mariotte de l'Académie Royale des Sciences*, 2 vols. (Paris, 1717), i, pp. 121-47.

[72] "Ma premiére hypothèse est, qu'il y a plusieurs principes grossiers & visibles des Plantes, comme l'eau, le soufre ou huile, le sel commun, la salpètre, le sel

In 1701 Homberg, a Dutch member of the *Académie des Sciences*, addressed the issue of the constituents of bodies. He produced a revised version of the chemical principles. Homberg recognised five principles as the result of chemical analysis: sulphur (active), salt, water and mercury (neutral) and earth (passive). He denied that mercury is a constituent of plants and animals, since it is found in minerals only. He criticised the accepted view of elemental and pure sulphur, as he claimed that pure sulphur is not obtained by chemical analysis. He speculated that it can be the matter of light. In addition he distinguished different kinds of sulphurs and of salts. When discussing the constituents of vegetables, Homberg placed special emphasis on the changes which the principles undergo within the body, rather than on their original properties. Corpuscular views played a relevant part in Homberg's chemistry and alchemy. He regarded gold and silver as composed of particles of sulphur and mercury, while other metals contained particles of other substances.[73]

ATOMISM AND CARTESIANISM IN THE NETHERLANDS

The combination of corpuscular theories and chemistry was very common among Dutch chemists active in the second half of the seventeenth century. Though aspects of Descartes's natural philosophy were adopted, the Cartesian system and his methodology were not. Dutch chemists stressed the importance of chemical practice and paid special attention to the teaching of chemistry. Their works show that they retained the chemical principles, but they reinterpreted them in corpuscular terms. Boyle's emphasis on the links between chemistry and corpuscular theories was generally supported, but his criticism of the status of the chemical principles was not. It was Hermann Boerhaave's task to develop Boyle's critique of

volatile ou armoniac, quelques terres, & c. Et que ces principes grossiers sont composés eux-mêmes de trois ou quatre principes plus simples, qui sont naturellement joints ensemble; par exemple, le salpétre a son flegme ou eau insipide, son esprit, son sel fixe, & c.; le sel comun a son flegme, son esprit, son sel fixe, & c. Et on peut croire ave beaucoup de vrai-semblance, que ces principes plus simples sont encore composés de quelques parties différentes entre elles, tellement petites, qu'on ne peut les appercevoir par aucun artifice, ni determiner quelles sont leurs figures & leurs autres proprietez.", ibid., p. 121.

[73] Homberg's chemical views are mainly contained in *Histoire et mémoires de l'Académie Royale des Sciences* (n. 66), 1702, 1704-6. On Wilhelm Homberg see Partington, iii, pp. 42-7, Holmes (n. 61), pp. 137-8; Bougard, *La Chimie* (n. 11), pp. 136-9; Sturdy, *Science* (n. 58), pp. 226-33.

the principles. Like Boyle, he maintained that the so-called principles did not pre-exist in the mixed body, but are produced by fire. Therefore, chemical analysis can hardly disclose the ultimate components of bodies. Moreover, according to Boerhaave, the substances obtained by analysis cannot recompose the mixed body they were extracted from. Finally, like Boyle and the majority of seventeenth-century Dutch chemists, Boerhaave regarded chemistry as an independent science.[74]

At the beginning of the seventeenth century atomism was promoted in the Low Countries by David van Goorle (Gorlaeus) (1591-1612) and by Isaac Beeckman (1588-1637). While Goorle's atomism was qualitative, Beeckman's has a clear mechanical orientation. Goorle's *Exercitationes Philosophicae* (1620) explicitly rejected the Aristotelian notion of matter and form and advocated the existence of indivisible units of matter as the ultimate constituent which remain unchanged in mixed bodies.[75] This does not mean that Goorle saw atoms as particles of a homogeneous matter. He retained two out of the four elements, namely water and earth, while he regarded fire as an accident.[76] Unlike the classical atomists, Goorle did not conceive atoms as particles endowed with only mechanical properties (size, shape, motion). He attributed to them various qualities, i.e., *humor* and *crassities*, *opacitas* and *diaphaneitas*.[77] The term atom is used by Goorle to mean both the ultimate (and simple) corpuscles of bodies and compound corpuscles ("ex atomis homogeneis facta sunt corpora homogenea, ex atomis heterogeneis heterogenea").[78] This shows that Goorle did not adopt the hierarchy of corpuscles, which may be found in Sennert and Basso. In Goorle's *Idea Physicae* there are occasional references to chemical topics. Goorle maintains that the *tria prima* account for the generation of metals

[74] For Boerhaave see DSB; H. Metzger, *Newton, Stahl, Boerhaave et la doctrine chimique* (Paris, 1930), pp. 191-305; G. Lindeboom, *Hermann Boerhaave, the Man and his Work* (London, 1968). See H. Boerhaave, *Elementa Chemiae*, 2 vols. (Leiden, 1732).

[75] David van Goorle, *Exercitationes Philosophicae quibus universa fere discutitur Philosophia Theoretica* (Leiden, 1620), pp. 235-251. On David van Goorle see Lasswitz, *Geschichte*, i, pp. 332-5 and 455-63 and T. Gregory, 'Studi sull'Atomismo del Seicento. II David van Goorle e Daniel Sennert', *Giornale critico della filosofia italiana* 45 (1966), 44-63.

[76] Goorle, *Exercitationes* (n. 75), pp. 313-4; 318.

[77] Ibid., pp. 143-4.

[78] Ibid., p. 247.

and stones. In addition, he explicitly states that the atomic theory of matter is not incompatible with the transmutation of metals.[79]

Beeckman's views of matter were never published. They were certainly known to some of his contemporaries, including Descartes, and were put down in his *Journal*, which came to light only in the twentieth century.[80] Beeckman adopted a mechanical theory of matter, stressing that explanations of natural phenomena must be based on the shape, motion and quantity of atoms. Beeckman's atoms are not indivisible in principle, but they cannot be physically divided as they do not have pores.[81] However, traces of the theory of elements may be found in Beeckman's *Journal*. He did not deny the existence of the so-called elements, namely homogeneous and simple bodies. Unlike the Aristotelian elements, Beeckman's consist of the same universal matter. To each element correspond different kinds of atoms, with distinct shapes.[82] Beeckman was reluctant to hypothesise the shape of atoms; rather he adopted explanations based on the different ways atoms are joined, producing clusters of atoms. As Kubbinga pointed out, Beeckman's matter theory contains a classification of corpuscles according to their complexity.[83] Beeckman called the primary clusters of atoms *homogenea*, which differ according to their textures. Both the four elements and the *tria prima* are defined as *homogenea* – each of them being composed of the same kind of atoms. Beeckman also introduced *homogenea* of compound substances. They differ according to the different ratio of the elements and principles and also according to their textures.[84] The notion of

[79] Goorle, *Idea Physicae* (Utrecht, 1651), pp. 51-2.

[80] C. de Waard (ed.), *Journal tenu par Isaac Beeckman*, 4 vols. (The Hague, 1939-53).

[81] Ibid., ii, pp. 245-6.

[82] "Videntur haec primo a materia prima primae differentiae constitui, ita ut non plures sint differentes figurae quam quatuor; ergo quatuor atomorum figurae constituunt quatuor differentias." Ibid., i, pp-152-3. "Atomi videntur tantum esse quatuor generum, quorum unum est ex quibus constat terra [...] ita ut pura terra constet ex solis atomis ejus generis." Ibid., iii, p. 138. As Kubbinga noted, Beeckman refrained from suggesting which were the shapes of the four kinds of atoms, see H.H. Kubbinga, 'Les premières théories «moléculaires»: Isaac Beeckman (1620) et Sébastien Basson (1621). Le concept d'«individu substantiel» et d'«espèce substantielle»', *Revue d'Histoire des Sciences* 37 (1984), 215-33, esp. p. 220.

[83] See H.H. Kubbinga, 'The first Molecular theory (1620): Isaac Beeckman (1588-1637)', *Journal of Molecular Structure* 181 (1988), 205-18.

[84] "Sufficiat dixisse elementorum minima in compositis non solum differre proportione numeri, ut 3 partes ignis, 4 aeris, 3 aquae, 5 terrae, sed etiam differre

homogenea is the foundation of Beeckman's definition of species and individual. Some *homogenea*, when they are joined in a given manner, constitute a species. The different quantity of *homogenea* produce the different individuals of the species.[85] As Kubbinga pointed out, "Son [Beeckman's] concept d' «homogeneum physique» implique d'ailleurs également la réduction des réaction chimiques à la disparition d'un ou plusieurs types d'«homogenea physiques» et l'apparition d'un ou plusieurs autres par suite d'un réarrangement des atomes".[86] When discussing chemical composition, Beeckman stressed the importance of the spatial arrangement of atoms (as well as of *homogenea*). The same kinds of atoms, in the same quantity, bring about different *homogenea*, and in turn the same simple *homogenea* produce different complex *homogenea* according to their spatial arrangement. Beeckman's emphasis on the texture, i.e., on the arrangement of corpuscles (both simple and composed) explains the specificity of substances (as well as the possibility of their transmutation) on the grounds of a limited number of constituents and provides a valid alternative to the substantial forms.[87]

Before the diffusion of Cartesianism, Dutch philosophy had a clear empirical orientation. This was due principally to two factors: the general tendency to stress the empiricist elements in Peripatetic philosophy; and the sudden influence of Francis Bacon, which affected the reception of Cartesian natural philosophy.[88]

Religious controversies had no little influence on the reception of Descartes's ideas in the Netherlands. As Theo Verbeek pointed out, though most Cartesians were orthodox and there was no logical connection between Descartes's thought and Remonstrantism, Cartesians were often regarded as linked to the Remonstrants. This circumstance involved Cartesianism in religious debates. As a result, eclecticism was very common among the early Cartesians, and relevant aspects of Descartes's metaphysics were often

situ et positione inter se. Sic alia proportio numeri et situs est in hominis venis, alia in nervis, alia in ossibus." Ibid., ii, p. 70.

[85] Ibid., ii, p. 128.

[86] Kubbinga, 'Premierès théories' (n. 82), p. 225.

[87] Besides the spatial arrangement of particles, the other cause of the change in texture is the ratio between particles and *vacuola*: "Unde colligitur varietatem rerum oriri ex proportionibus vacui et corporis." Beeckman, *Journal* (n. 80), ii, p. 238. See also ibid., iii, p. 56.

[88] See P. Dibon, *La Philosophie Néerlandaise au siècle d'or* (Amsterdam, 1954), p. 206.

rejected. This is borne out, among others, by Adriaan Heereboord (1614-61), Johannes de Raey (1622-1707) and Johannes Clauberg (1622-65).[89]

At the University of Utrecht Cartesianism was supported by Henricus Regius, Professor of Theoretical Medicine and Botany, who defended Descartes's philosophy in 1641. This occurred in medical disputations, where Descartes's mechanical principles were unambiguously adopted. Philosophical issues were also relevant in Regius's disputations. Matter is defined as extension or quantity and the particulate theory of matter is explicitly adopted.[90] Regius was attacked by Gysbertus Voetius, a minister of the Utrecht Church, and by the theological faculty. The controversy between Voetius and Regius finished in 1642 when the University of Utrecht condemned Descartes's philosophy. Later, in 1654-5, Cartesianism was accepted at Utrecht by Johannes de Bruyn (1620-75) and became part of the University teachings. It is to be noticed that Cartesian metaphysics was often dropped by Utrecht professors.[91] Like Descartes, Regius held a fully mechanical theory. However, his philosophy diverged from Descartes's at two points. Regius rejected Descartes's dualism of body and soul and held a materialistic view of the human body. The other difference is to be found in his *Philosophia naturalis* of 1661, where Regius admits the existence of atoms, which he had denied in the *Fundamenta* of 1646.[92] Chemistry plays little part in Regius's natural philosophy and chemical phenomena are explained by means of the mechanical properties of corpuscles, namely their sizes and motions.[93]

[89] T. Verbeek, *Descartes and the Dutch. Early Reactions to Cartesian Philosophy, 1637-1650* (Carbondale and Edwardsville, 1992), pp. 5-9. See also Dibon, *La Philosophie Néerlandaise* (n. 88), *passim*.

[90] Verbeek, *Descartes and the Dutch* (n. 89), pp. 13-17.

[91] Ibid., pp. 17-33.

[92] See P. Farina, 'Il corpuscolarismo di Henricus Regius: materialismo e medicina in un cartesiano olandese del seicento', in *Ricerche sull'atomismo del seicento* (Florence, 1977), pp. 119-78. Regius's *Philosophia naturalis* was a new edition with small changes of the *Fundamenta*. It was first published in 1654 and again in 1661.

[93] See for instance the precipitation of mercury: "Mercurius, & omnia metalla in aquis fortibus soluta, in iisque, propter parvitatem dissolutarum particularum, & vehementem aquarum istarum motum, volitantia, adminiculo salis vel calcis tartari injectae, sub specie pulveris ad fundum praecipitantur; quia sal vel calx tartari habet particulas ita conformatas, ut vi ebullitionis, quam excitat, praecipuos spiritus ex aqua forti expellat; & deinde, ut metallorum particulis & aliis salibus, in aqua forti exsistentibus, facile adhaerat, & multas inter se conjungat, quo illae graviores factae,

The introduction of Cartesianism in Leiden was due mainly to Heereboord, Professor of Logic.[94] From 1644 Heereboord embraced Cartesianism, and, despite the decree of 1647 that prohibited the teaching of Cartesianism, he included Descartes's philosophy in his disputations. Though he upheld Descartes's method, Heereboord can hardly be styled an orthodox Cartesian. His *Disputatio philosophica de atomis* (Leiden, 1650) shows a favourable attitude towards atomism.[95] Heereboord's version of atomism is not purely mechanical, since his atoms are endowed with qualities – as attested by the reference to atoms of the elements and of the chemical principles.[96]

A friend of Descartes, the Leiden physician Cornelis van Hoogelande adopted the French philosopher's mechanism in medicine. Like Regius, Hoogelande dropped Descartes's dualism and gave a materialistic solution to the soul/body problem. The soul is the same as the *materia subtilis*, which can penetrate all bodies.[97] Unlike Descartes, Hoogelande awards special importance to chemistry in physiology. Life is conceived as the outcome of fermentation which occurs in blood. Hoogelande defines fermentation along corpuscular lines, namely as a slow movement of particles of matter, and he compares it to the production of heat when butter of antimony and spirit of nitre are mixed.[98]

& minus fortiter motae, quam ut a liquore isto possit sustineri, necessario ad fundum subsidunt." H. Regius, *Fundamenta physices* (Amsterdam, 1646), p. 128. See S. Matton, 'Cartésianisme et Alchimie: à propos d'un témoignage ignoré sur les travaux alchimiques de Descartes. Avec une note sur Descartes et Gómez Pereira', in Greiner (ed.), *Aspects* (n. 2), pp. 111-84, esp. pp. 124-5.

[94] Verbeek, *Descartes and the Dutch* (n. 89), pp. 34-40.

[95] For Heereboord see also Dibon, *La Philosophie Néerlandaise* (n. 88), pp. 116-9.

[96] A. Heereboord, *Meletemata Philosophica* (Leiden, 1659), pp. 347-51.

[97] C. Hoogelande, *Cogitationes* (Amsterdam, 1646), pp. 26-8. For Hoogelande see S. Matton, 'Cartésianisme et Alchimie: à propos d'un témoignage ignoré sur les travaux alchimiques de Descartes. Avec une note sur Descartes et Gómez Pereira', in Greiner (ed.), *Aspects* (n. 2), pp. 118-23.

[98] Hoogelande, *Cogitationes* (n. 97), p. 79: "Fermentatio autem generaliter a nobis definienda videtur, languidior ac moderatior (velocior enim, ac vehementior, effervescentia dici solet) materiae humidae vel liquidae, vel variarum materiarum commixturae, actio, tamquam tertium quid, sive tertius quidam motus: vel tepidioris externi caloris adminiculo intercedente; vel solius compositionis vel commistionis ratione, ex diversitate motus interni ac insensibili insensibilium particularum a diversitate, tum quantitatis & qualitatis earundem, tum pororum ipsarum

Atomism was supported at Franeker by Jan Fokkesz (1618-51), who was known as Johannes Phocylides Holwarda. Physician, astronomer and philosopher, Holwarda kept the Aristotelian terminology, but unambiguously founded natural philosophy on the corpuscular theory of matter. His *Philosophia Naturalis, seu Physica Vetus-Nova*, published posthumously in 1651, starts with the definition of matter and form. Both notions are reinterpreted by Holwarda along corpuscular lines. Matter is extended and divided into atoms; form is the texture of atoms.[99] Bodies are formed of atoms and void. Atoms are solid corpuscles which receive motion directly from God. The elements do not disappear from Holwarda's philosophy, but they are reinterpreted in corpuscular terms. The atoms of earth are cubic, those of water long and flexible, those of air have the same shape as those of water, but are smaller; finally, those of fire are spherical. Like Sennert and Basso, Holwarda distinguished different kinds of atoms, some simple, others compound.[100] As all natural phenomena are produced by physical agents, sympathy and antipathy are explained as the outcome of the proportion and dispositions of atoms.[101]

DUTCH CHEMISTS

Sylvius's teaching in Leiden gave strong impetus to chemistry in the Netherlands. Another important contribution to the development of chemistry came from the work of the German chemist Johann Rudolph Glauber, who founded a laboratory of repute in Amsterdam where several chemists were trained. Since 1660 chemical laboratories and courses had spread in various Dutch cities. Carolus Ludovicus van Maets (1640-90) ran a chemical laboratory and lectured at Utrecht. From 1694 chemistry was taught at Utrecht by Johann Conrad Barckhausen (1666-1723). Jacob Le Mort had a laboratory and taught chemistry in Leiden. At the end of the seventeenth century most Dutch physicians were iatrochemists, and some of them adopted corpuscular theories of matter.

substantiarum prodeunte, orta, qua mediante, humidiori fermentandae vel fermentatae substantiae portioni, subtilioris aëris quantitas, vel materiam quaedam aetherea involvitur." See also ibid., p. 81.

[99] J. Phocylides Holwarda, *Philosophia Naturalis, seu Physica Vetus-Nova* (Franeker, 1651), pp. 7-8. For Holwarda see Dibon, *La Philosophie Néerlandaise* (n. 88), pp. 155-7.

[100] Holwarda, *Philosophia* (n. 99), p. 15.

[101] Ibid., pp. 16-17.

Fermentation plays a central role in the iatrochemistry of Franciscus de le Boë (Sylvius) (1614-1672). He built his own theories on the chemical principles and gave prominence to the acid/alkali reactions. Though he occasionally referred to the motion of particles, there is no evidence that Sylvius tried to combine iatrochemistry and corpuscular philosophy.[102] Such a combination became common among Dutch physicians and chemists in the second half of the seventeenth century.

Carel de Maets (1640-90), who received his training from Glauber in Amsterdam, lectured in Utrecht and then was appointed Professor of Chemistry at Leiden in 1669, where a laboratory had been founded in 1665. He taught chemistry until his death in 1690. In 1694 the teacher of chemistry in Leiden became Jacob Le Mort (1650-1718), who had started his career as a teacher of theology in 1664 and had then worked in Glauber's laboratory in Amsterdam. At Leiden he attended de Maets's lectures and taught privately. After de Maets died, Le Mort was appointed supervisor of the laboratory in 1695, and Professor in 1702.[103] Jacob Le Mort based chemistry on the mechanical philosophy, namely on matter and motion. Following Descartes, Le Mort states that matter is extended and divisible. His adherence to the Cartesian mechanism does not bring about the rejection of the chemical principles. They are reinterpreted in corpuscular terms. For Le Mort, the so called principles are not to be thought of as simple substances. All bodies are composed of particles which are in movement or at rest. Two kinds of corpuscles, one fluid the other fixed, account for the variety of natural substances including the chemical principles.[104] Le Mort's

[102] A good account of Sylvius's iatrochemistry is Partington, ii, pp. 281-90. See also DSB. For Sylvius's theory of fermentation see Sylvius, *Disputationum Medicarum*, in *Opera Medica* (Amsterdam, 1679), pp. 10-13. Sylvius's doctrine of fermentation was first presented in a disputation of 1659. See *Disputatio prima De Alimentorum Fermentatione in Ventriculo* (Amsterdam, 1659), repr. in *Opera Medica*, p. 11: "Mutationem, quam in ventriculo subeunt alimenta, quamque impraesentiarum examinare fert animus, Chylificationis nomine vulgo indigitant; nobis Fermentationis nomen magis arridet ob rationes mox secuturas. Utique duplex mistorum observatur destructio & dissolutio: Altera quidem violenta & subito cum notabili partium dissipatione contingens ab Igne, Ustio dicta; altera vero blanda & lente citra notabilem partium iacturam contingens per Aquam, Fermentatio, vel quando faetor coincidit, Putrefactio vocata."

[103] On Jacob Le Mort, see Partington, ii, pp. 737-8.

[104] "Ex hisce duobus principiis, fluido nempe & firmo oriuntur sequentia principia chymicorum, quae quamvis videantur externis nostris sensibus, inter se invicem differre, attamen in haec duo, facili negotio concurrunt, & reduci possunt,

principles are salt, water, earth and a spiritual or ethereal substance, which he identifies with the chemists' mercury. They differ in the shape, size and motion of their corpuscles.[105] Despite the strong mechanical orientation of his work, Le Mort follows the chemical philosophers in the explanation of the origin of qualities. In his view, salt is responsible for the production of odours, tastes and colours.[106] Like Boyle, Le Mort places special emphasis on the notion of texture in the explanation of chemical phenomena.[107] Though Le Mort's theory of matter is strictly mechanical he did not deny chemistry its autonomous role in the investigation of natural phenomena. Quoting Boyle, Le Mort stresses that chemistry explains a variety of phenomena: "Unde latissimus resultat campus causarum & effectuum corporum naturalium indagandorum per Chymiam."[108]

Boyle's criticism of the chemical principles is shared by Wilhelm ten Rhyne (1647-1700), physician of the East India Company, who adopted the corpuscular theory.[109] Unlike Boyle, ten Rhyne sets out speculations on the shapes of corpuscles. He tries to explain the properties of fire, water and salt as the direct consequence of the shapes of their constituent corpuscles. Fire corpuscles are pointed, while those of water are like cylinders, and those of salt have different shapes, but all are endowed with a sharp point. Ten Rhyne is representative of the common mixture of corpuscularianism and chemical theories. An eclectic scientist, he embraces much of Descartes's philosophy, in particular the *materia subtilis* and the rejection of vacuum. However, he adopts atoms but not the extended divisible matter. Along with

nempe in spirituosam & aëream aquam & terram spongiosam." Le Mort, *Compendium Chymicum, demonstrans Experimentis & Rationibus brevem & facilem Methodum Operationes accurate & succinte ad finem producendi* (Leiden, 1682), p. 5.

[105] Ibid., pp. 6-10. In his *Idea Actionis Corporum* (Leiden, 1693), pp. 6-7, Le Mort defines the chemical principles as follows: "Primo puncta minima rigida, acuta, quae salia vocantur, secundo oblongae, molles, obtusae & flexiles particulae, sub nomine aquae sese offerentes. Tertio corpora dura, solida, in omnem dimensionem valde extensa, ad motum per se inertia quae terrae vocantur. Ad haec tria omnia reduci posse corpora, eorundemque actiones & figuras."

[106] "Quod attinet colores, odores, sapores illos a particulis salinis dependere statuimus." Ibid., p. 10.

[107] Ibid., pp. 37-8.

[108] Le Mort, *Chymiae Verae Nobilitas & Utilitas* (Leiden, 1696), p. 3.

[109] See W. ten Rhyne, *Exercitatio Physiologica* (Leiden, 1669), sig. B4^r. For ten Rhyne see A. Hirsch, *Biographisches Lexikon der hervorragenden Ärzte* 6 vols. (Vienna and Leipzig, 1885-7).

Boyle, he stresses the importance of the texture of corpuscles in the explanation of natural phenomena.[110]

Atomism was adopted by J.C. Barckhausen in his chemical textbook. He also endorsed the chemical principles, which he divided into three groups: active (salt and oil), neutral (water) and passive (earth).[111] The principles are conceived as the simplest and smallest particles of bodies. The corpuscular reinterpretation of the chemical principles may also be found in Barckausen's subsequent works, i.e., *Compendium Ratiocinii Chemici* (1712) and *Elementa Chemiae* (1718).

The reduction of chemistry to physics, i.e. to the laws of mechanics, is pursued by the Dutch physicist Nicolas Hartsoeker (1656-1725), who lived many years in France and in 1699 became a member of the *Académie des Sciences*. He modified the Cartesian theory of principles by assuming two elements: one an infinite fluid, a continuum, whose particles are always in motion; the other solid, i.e. an infinite number of hard corpuscles of different sizes and shapes, floating in the first element.[112] The physical and chemical properties of bodies are explained by means of the shape of the particles. Hence fluid bodies are composed of spherical or oval particles. Water particles are hollow globes; air particles are hollow and elastic. Cohesion is produced by the pressure of the first element, and weight by the impact of corpuscles. Taste is explained by the shapes of corpuscles.[113]

In his subsequent work (*Conjectures Physiques,* 1706) Hartsoeker adopts a probabilistic approach to science, stressing that all natural phenomena ultimately depend on God. In this work he presents his own theories as mere conjectures. It is impossible to establish the exact shape and size of the corpuscles, but it is both legitimate and useful to make hypotheses about them in order to give plausible explanations of natural phenomena. Hartsoeker's probabilism is perfectly compatible with the general mechanist view, already presented in his *Principes de Physique* (1696). In the *Conjectures* chemistry receives more attention than in the previous work. Hartsoeker describes the different chemical substances and their properties

[110] W. ten Rhyne, *Meditationes in Magni Hippocratis textum XXIV De Veteri Medicina* (Leiden, 1672), pp. 70; 72; 246-53; 303-4; 361.

[111] J.C. Barckhausen, *Pyrosophia* (Leiden, 1698), pp. 7; 13-15; 33. On Barckhausen (or Barchusen), see Partington, ii, pp. 700-2 and F. Abbri, *Le Terre, l'Acqua, le Arie. La Rivoluzione chimica del Settecento* (Bologna, 1984), p. 22.

[112] N. Hartsoeker, *Principes de Physique* (Paris, 1696), pp. 1-2. On Hartsoeker, see DSB and Partington, ii, pp. 451-4.

[113] Hartsoeker, *Principes de Physique* (n. 112), pp. 88-95; 99-102.

as the outcome of their geometrical forms, their sizes and weights. Acid salts are made of long and sharp particles, which, when mixed with water, form acid spirits. The latter are described as balls studded with sharp points. Hence they can easily penetrate the texture of bodies. Alkali have hollow cylindrical particles, where acids' points can enter.[114] The long series of mechanism used by Hartsoeker shows his strong faith in the mechanical explanation and very little interest in the investigation of chemical phenomena. His approach is deductive and highly speculative. Hartsoeker never pursued independent empirical testing for his theories. His reductionist approach marks a difference from Boyle's chemistry – which, as we have seen, is based on the specific chemical properties of bodies.

CHEMISTRY AND ATOMISM IN GERMANY

In the second half of the seventeenth century chemistry had achieved a prominent position in German states – in both the universities and in the courts. The institutionalisation of chemistry was produced by two main factors: the German princes' support of alchemy and Paracelsianism; and the exploitation of mines. Paracelsianism was generally tolerated in universities, and the establishment of new universities (twenty-two in the seventeenth century) contributed to the development of iatrochemistry in the medical faculties. The first chair of chemical medicine (*chymiatria*) was established at Marburg in 1609, though it is likely that Libavius lectured on chemistry from 1592. After Marburg, chemistry was taught at the University of Jena, where Rolfinck was Professor of Chemistry from 1641 to 1673. He was succeeded by George Wolfgang Wedel, Friedrich Hoffmann and then George E. Stahl. M. Ettmüller and J. Bohn taught at Leipzig, where the latter became Dean of the Medical Faculty; Johann Kunckel taught at Wittenberg, where he established a laboratory in which he discovered the method of making phosphorus. Johann Andreas Stisser was Professor of Chemistry at Helmstädt and Friedrich Hoffmann taught chemistry at Halle for three decades. The *Academia Curiosorum* paid special attention to chemistry, witness the Academy's periodical, the *Miscellanea* or *Ephemerides*. Many chemical laboratories were created in the seventeenth century, the most important were at Munich (Becher), at Dresden and Berlin (Kunckel), at Altdorf (J.M. Hoffmann), and at Jena (Rolfinck).[115]

[114] Hartsoeker, *Conjectures Physiques* (Amsterdam, 1706), pp. 101-130.

[115] See B.T. Moran, *Chemical Pharmacy Enters the Universities: Johannes Hartmann and the Didactic Care of "Chymiatria" in the Early Seventeenth Century*

Throughout the seventeenth century German chemistry was closely connected with medicine and pharmacy. J-B. van Helmont's ideas had a strong and lasting impact on German chemists, on both those who lived in the German states and on those (like Polemann, Clodius, Glauber and Tachenius) who settled in other countries. The most important and influential German chemist in the mid-seventeenth century was Johann Rudolph Glauber, who spent many years in the Netherlands and established laboratories both in Germany (Kissingen) and in Amsterdam.[116] Though Glauber was essentially a practical chemist, his works contain some theories, often expressed in a rather unsystematic way, on the principles of natural bodies. Glauber accepted the Paracelsian principles and reinterpreted the spirit of the world as the universal salt. Glauber's identification of the universal salt with nitre has been used to stress the importance of Glauber's experimental achievements in chemistry. Nevertheless, as is apparent from his tract on salt, he based his views on both the Scriptures and experiments.[117] Besides the universal salt, the notion of *semina* plays a central part in Glauber's chemistry, in particular in his explanations of the origin and growth of minerals and metals. *Semina* originate in the stars and are carried into the bowels of the earth by air. Earth acts as the matrix, where metals are produced from specific seeds. Glauber often extolled van Helmont and believed in the *Alkahest*, but he, unlike the Belgian physician, made no use of corpuscular theories in his chemical work, which has much in common with Paracelsus's.

After Sennert and Sperling, the corpuscular theory of matter was developed by Joachim Jungius (1587-1657), whose theory of matter has been thoroughly investigated by H. Kangro and C. Meinel.[118]

(Madison, Wisc., 1991); id., *The Alchemical World of the German Court. Occult Philosophy and Chemical Medicine in the Circle of Moritz of Hessen* (Stuttgart, 1991), pp. 50-67; H. Trevor-Roper, 'The Court Physician and Paracelsianism', in V. Nutton (ed.), *Medicine at the Courts of Europe. 1500-1837* (London and New York, 1990), pp. 79-94. Useful information on German chemistry in the late seventeenth century may be found in K. Hufbauer, *The Formation of the German Chemical Community (1720-1795)* (Berkeley, 1982).

[116] On Glauber see DSB; Partington, ii, pp. 341-61; Debus, *The Chemical Philosophy*, 2 vols. (New York, 1977), pp. 425-41.

[117] See R. Glauber, *Tractatus de natura salium* (Amsterdam, 1659).

[118] H.H. Kangro, *Joachim Jungius' Experimente und Gedanken zur Begründung der Chemie als Wissenschaft* (Wiesbaden, 1968) and C. Meinel, 'Der Begriff des chemischen Elementes bei Joachim Jungius', *Sudhoffs Archiv* 66/4 (1982), 313-38.

Despite the support Paracelsianism received from German princes, and the establishment of chemistry as part of the curricula in many universities, Aristotelianism kept a strong hold in academic philosophy and medicine – at least until the end of the seventeenth century. Attacks on Paracelsianism – notably on the most innovative and radical doctrines – were launched by Anton Günther Billich and Hermann Conring. Nevertheless, they did not deny the importance of chemistry, but insisted that it was a practical art, having nothing to say about philosophical issues such as the principles of natural bodies. Their strictures on chemistry were later reassessed by Werner Rolfinck.[119]

The diffusion of Cartesianism in German universities was by no means easy, born out by the fact that it was prohibited in the Calvinist Universities of Herborn and Marburg. However, Cartesian philosophy was accepted by a number of philosophers and physicians (some of them being trained in the Netherlands), but it was often adapted within the context of scholastic metaphysics.[120] The rapid diffusion of Boyle's corpuscular philosophy might be regarded as the outcome of the restrictions imposed on Cartesianism. Boyle's corpuscular views were often introduced in the context of experimental works, not as a new system of natural philosophy. Both in philosophy and in medicine syncretism was widely adopted in Germany.[121] Cartesianism was often reconciled with Scholasticism and the corpuscular theories of matter coexisted with iatrochemical doctrines – notably those of

[119] A.G. Billich, *Thessalus in Chymicis Redivivus, id est, de vanitate medicinae chymicae seu spagyricae, dissertatio. Eiusdem anatomia fermentationis* (Frankfurt, 1640). Ferguson refers to a 1639 edition which I have not been able to find; H. Conring, *De Hermetica Ægyptiorum Vetere et Paracelsicorum Nova Medicina* (Helmstaedt, 1648) and W. Rolfinck, *Chimia in artis forma redacta* (Jena, 1661).

[120] F. Trevisani, *Descartes in Germania. La ricezione del cartesianesimo nella Facoltà filosofica e medica di Duisburg (1652-1703)* (Milan, 1992), pp. 15-17.

[121] On seventeenth-century German philosophy see especially M. Wundt, *Die deutsche Schulmetaphysik des 17. Jahrhunderts* (Tübingen, 1939); L.W. Becke, *Early German Philosophy* (Bristol, 1996^2), pp. 160-95, and S. Wollgast, *Philosophie in Deutschland zwischen Reformation und Aufklärung* (Berlin, 1988). It is apparent that Boyle's philosophical views attracted the interest of German philosophers, as attested by the debate on the notion of nature, which originated from Boyle's *Free Enquiry into the Vulgarly Received Notion of nature* (1686) and involved G.C. Schelhammer, J.C. Sturmius and Leibniz. See G.C. Schelhammer, *Natura sibi et Medicis Vindicata, sive de Natura...* (Kiel, 1697); J.C. Sturmius (*praeses*), *Exercitatio Philosophica de Natura Sibi Incassum Vindicata* (Nordlingen, 1698) and G.W. Leibniz, *De Ipsa Natura* (1698).

van Helmont and Sylvius. This was also the case with the reception of Boyle's theory of matter. Boyle's corpuscular chemistry was generally accepted by German iatrochemists, who often stressed the continuity between Boyle's and van Helmont's chemistry.

Members of the *Academia Naturae Curiosorum* (founded in 1652) paid special attention to Boyle's works, as attested by the *Miscellanea Curiosa*.[122] The *Acta Eruditorum* contain many reviews of Boyle's works. Finally, Daniel Georg Morhof praised Boyle's corpuscular philosophy in his works and translated a number of Boyle's tracts into Latin.[123]

BECHER'S PRINCIPLES

Though Johann Joachim Becher's chemical works were rather popular at the beginning of the eighteenth century, they have not received adequate attention from science historians.[124] In the present section I am focusing on Becher's theory of principles, which includes a reinterpretation of the Paracelsian principles. Corpuscular views were also adopted by Becher though he never accepted the reduction of chemistry to the mechanical theory of matter.

[122] The publication of *Miscellanea Curiosa, sive Ephemeridum Medico-Physicarum Germanicarum Academiae Naturae Curiosorum* started in 1670. On the *Academia Naturae Curiosorum* see J.W. Evans, 'Learned Societies in Germany in the Seventeenth Century', *European Studies Review* 7 (1977), 129-151. Chemistry and medicine played a prominent part in *Miscellanea Curiosa*: see the contributions, among others, of Philip J. Sachs, J. Langelot, Heinrich Screta and G. Wolfgang Wedel.

[123] See C.W.T. Blackwell, 'Sturm, Morhof and Brucker vs. Aristotle: three eclectic natural philosophers view the Aristotelian method', in D.A. Di Liscia, E. Kessler and C. Methuen (eds.), *Method and Order in Renaissance Philosophy of Nature* (Aldershot, 1997), pp. 381-407, esp. pp. 388-9.

[124] The best account of Becher's chemistry is still Partington, ii, pp. 637-652, though he paid little attention to Becher's sources. See also J. Berger, *Ideen über die Verwandlung der Stoffe. Chemische Materietheorien und Affinität im 17. und 18. Jahrhundert* (Berlin, 1998), pp. 25-33. For biographical sketch, see DSB and M. Teich, 'Interdisciplinarity in J.J. Becher's thought', *Journal of European Ideas* 9, (1988), 145-160. On Becher's views of politics and economics, see G. Frühsorge and G.F. Strasser (eds.), *Johann Joachim Becher* (Wiesbaden, 1993), and P.H. Smith, *The Business of Alchemy. Science and Culture in the Holy Roman Empire* (Princeton, 1994). For Becher's alchemy see Principe, *Aspiring Adept* (n. 11), pp. 112-3; 173-4.

In the *Oedipus Chimicus* Becher adopted the three chemical principles and put special emphasis on the notion of *semina*: "quare concludo, tum semen esse veram primam Chymicorum materiam, cum generandi & vegetandi integram potentiam habet."[125] He identified the remote matter of metals with a mercurial principle, which he called "aqua gravissima ex mercuriali vapore orta." Becher articulated his theory of principles in the *Physica Subterranea* of 1669, where he maintained that earth and water are the two material principles, while air and heaven are the formal ones (*miscentia*).[126] In Becher's view, the true material principle is earth, which he divided into three species: vitrifiable earth (*terra prima or terra vitrescibilis*), the principle of fixity and solidity; oily earth (*terra pinguis*), the principle of combustibility; fluid earth (*terra fluida*), the principle of volatility. Following Boyle, Becher rejected the Paracelsian theory of principles by stating that they were by no means simple substances but mixed bodies.[127] It is however apparent that his three types of earths played a role which was analogous to the *tria prima*. With their combinations (and by mixing with water) the three earths produce different substances, as well as the qualities of natural of bodies.[128]

[125] J.J. Becher, *Oedipus Chimicus* (Frankfurt, 1664), p. 18.

[126] J.J. Becher, *Actorum Laboratorii Chymici Monacensis, seu Physicae subterraneae libri duo* (Frankfurt, 1669), pp. 42-8.

[127] "Neminem autem spero, ita absurdum fore, ut praefata tria principia aliter quam propinqua & principiata intelligat, nempe pro materia jam proxime ad actum disposita, & licet hoc modo considerentur, tamen quomodocunque explicentur, improprie sal, sulphur & Mercurius dicuntur, si enim sulphur commune, seu qualecumque intelligas, illud mixtum erit, & hoc de Mercurio & sali communi intelligendum, quae omnia mixta sunt, & ex partibus constant, quae definiuntur, quod debeant esse simplicia, haec vero Paracelsi Principia, non modo mixta, sed & decomposita sunt." *Actorum* (n. 126), p. 124. Boyle's anti-Paracelsian arguments are referred to in *Actorum*, pp. 457-8.

[128] "Nonnulli credunt, omnia constare ex sale, sulphure et Mercurio; sed ego probabo, omnia, seu potissima mixta, constare ex triplici terra, una vitrescibili, quae salis vicem praebat, matricem et basin, altera pingui, quae sulphur est, compagem, tincturam et tenacitatem dat, tertia subtilis est, et materiam supplet, Mercurius vocatur seu potius Arsenicum. Prima terra dat corpus ac substantiam et hypostasin mixtis, et est duplicis generis; vel calcinabilis, vel vitrescibilis; unde in animalibus ossa, in vegetabilibus cineris elixati, in mineralibus lapides. Secunda terra dat mixtis consistentiam, colorem, saporem etc. et est duplicis generis; consistens vel liquida; unde in animalibus sevum, adeps, axungia; in vegetabilibus oleum, gummi; in mineralibus et metallis sulphur, bitumen. Tertia terra dat mixtis formam,

An interesting aspect of Becher's theory of matter is his classification of substances according to their increasing complexity – which originated with Sennert and was subsequently adopted by Boyle. The classification started with simple substances: *simplicia*, then *composita*, *decomposita* and finally *superdecomposita*.[129] Though Becher never articulated a corpuscular theory of matter, he explained a number of chemical reactions in terms of particles. The increase of weight of metals on calcination is explained with the addition of igneous corpuscles.[130] Mineral waters are produced by particles of copper and iron.[131] Fermentation requires the action of air, whose particles dilate and rarefy the fermenting matter.[132] The properties of van Helmont's *Alkahest* are described along corpuscular lines: it analyses bodies to their smallest atoms ("in tenuissimas atomos").[133] Evidently Becher's particles have chemical, not mechanical properties. Becher's matter is not inert, but endowed with powers. The ultimate cause of motion is a spiritual principle.[134] Substances have mutual attractions, as silver and nitre bear witness.[135]

GERMAN IATROCHEMISTS AND HELMONTIANS

Helmontian doctrines are adopted by numerous chemists and physicians active in the second half of the seventeenth century. Their adherence to Helmontian ideas is often rather selective. The Helmontian principles (seeds and water) are accepted – and often reinterpreted in corpuscular terms –

penetrantiam, odorem, pondus, splendorem, lucem, etc. Est quoque duplicis generis, vel pura et tum est terra, vel mixta et tum est salina, in animalibus eam cernimus in eorum salibus volatilibus; in vegetabilibus in illorum aquis destillatis, spiritibus et aquis ardentibus in fuligine; in mineralibus conspicimus eam vel fluidam, ut in argento vivo, vel consistentem, ut in arsenico." Becher, *Alphabetum Minerale, in Tripus Hermeticus Fatidicus* (Frankfurt, 1689), pp. 105-7.

[129] Becher, *Actorum* (n. 126), p. 525.

[130] Ibid., p. 194.

[131] Ibid., pp. 96-7.

[132] Ibid., pp. 348 and 353.

[133] "Noster vero liquor non alia ratione solvat, quam penetrando, & corpora in tenuissimas Atomos redigendo." Ibid., p. 176.

[134] Ibid., pp. 12-14.

[135] "Aurum communem Mercurium appetit, argentum nitrum amat, quod in nitro sulphur seu terra secunda sit qua argentum ad perfectionem auri indiget, hinc fit, ut singulis solutionibus argenti in spiritu nitri, semper nonnihil auri inveniatur, quare argentum eadem ratione nitrum appetit." Ibid., pp. 400-1.

while the *Archeus* is often dropped. An exception was Johannes Dolaeus, who adopted van Helmont's notion of *Archeus* in his *Encyclopaedia*.[136] Dolaeus distinguished two different *Archei*, namely, one in the stomach and one in the heart. Pathological processes are generated by the alteration of the *Archeus*. Dolaeus's *Encyclopaedia* was a compilation, where one can find Paracelsian and Helmontian views together with theories of Descartes, Sylvius and Willis. The syncretistic approach is particularly apparent in the fourth book, dealing with fevers.

In Michael Ettmüller's *Medicina Hyppocratis Chymica* the Aristotelian elements and the Paracelsian chemical principles are rejected by means of arguments borrowed chiefly from van Helmont and from Robert Boyle. Following Boyle's views as expressed in *The Origine of Formes and Qualities*, Ettmüller rejected the Aristotelian and the Paracelsian doctrines of qualities too. These are generated, he maintained, by the various motions and contextures of particles.[137] The acid/alkali theory occupies a large part of the theoretical section of *Chimia Rationalis ac Experimentalis Curiosa* (published posthumously, possibly compiled from student notes).[138] Besides acid and alkali, Ettmüller admitted the existence of a third salt, i.e. the neutral salt.[139] He also intervened in the discussion on the nature of vegetable salts, by stating that fixed salts extracted from vegetables were artificial and did not pre-exist.[140] Corpuscular theories are often adopted in *Chimia Rationalis*, notably in the explanation of effervescence and fermentation, which he saw as the outcome of a rapid motion of particles.[141] Chemical combination and precipitation are produced by the weight and pores of the constituents' corpuscles.[142]

[136] J. Dolaeus, *Encyclopaedia Medicinae Theoretico-Practicae*...(Frankfurt, 1684). On Dolaeus (1638-1707), physician to the Landgrave of Hassen-Cassel, and a member of the *Academia Naturae Curiosorum*, see Hirsch (n. 109).

[137] M. Ettmüllcr, *Mcdicina Hyppocratis Chymica* (Leiden, 1671), pp. 14 8 and 43-4. (first edn Leipzig, 1670). Ettmüller (1644-1683) was Professor of Botany and Medicine at Leipzig, where he also lectured on chemistry, and a member of the *Academia Naturae Curiosorum*. See Partington, ii, pp. 298-300.

[138] Ettmüller, *Chimia Rationalis ac Experimentalis Curiosa* (Leiden, 1684).

[139] Ibid., p. 7.

[140] Ibid., p. 6.

[141] Ibid., pp. 43-5.

[142] Ibid., pp. 46 and 49: "praecipitatio est separatio corporis soluti. Ratio quapropter denuo separatur seu praecipitatur corpus a solvente est duplex: primo praecipitatio sit propter porulorum seu spaciolorum in menstruo solventem angustiam, ut particulae non amplius possint contineri in iisdem, secundo,

Van Helmont's theory of seeds and water was adopted in 1672 by David von der Becke, who accepted Tachenius's acid/alkali doctrine too.[143] He also maintained that fixed salts did not pre-exist in mixed bodies, but were produced by fire, as terrestrial particles were joined to volatile salts.[144] In his subsequent work (1674), devoted to the principles of natural bodies, von der Becke developed a corpuscular interpretation of the Helmontian doctrines.[145] The chemical principles, according to him, were not the ultimate constituents of bodies. Sulphur, as well as spirits, can be reduced into "alias particulas priores." The same is possible with salts: "salia omnia in aquam elementalem reducibilia esse."[146] The formal principle being the seed, earth is the 'matrix', water the material principle.[147] Similar views can be found in the chemical work of Johann Helfrich Jungken (1648-1726), *Physicus Ordinarius* in Frankfurt.[148] Jungken reinterpreted Tachenius's principles in corpuscular terms: acids are described as "corpuscula luminosa penetrantissima & subtilissima", their particles are flexible and are always in movement. Alkali are composed "ex particulis solidis, ponderosis, oblongis & rectis, inflexibilibus tamen & rigidis", having a texture which facilitates their combination with acid particles.[149] Jungken imposed severe

paecipitatio fit propter gravitatem particularum solutarum, ut non amplius in liquore pendere aut revolutare queant."

[143] D. von der Becke, *Epistola ad Praecellentissimum Virum Joelem Langellottum qua Salis Tartaris* (Hamburg, 1672), p. 12. On David von der Becke (1648-84), physician from Minden, see Hirsch (n. 109). An extract of von der Becke's letter to Langellott was published in the *Philosophical Transactions*, 1673, pp. 5185-93.

[144] Von der Becke, *Epistola* (n. 143), pp. 13-14.

[145] Von der Becke, *Experimenta et Meditationes circa Naturalium Rerum Principia* (Hamburg, 1683, first edn 1674). This work was reviewed in *Philosophical Transactions*, 1674, pp. 60-4.

[146] Von der Becke, *Experimenta et Meditationes* (n. 145), pp. 21; 25.

[147] "Terra est matrix seminum, in qua semina explicentur & nutriantur. Quod quidem cunctorum seminorum nutrimentum est Aqua, ab acido seminum fermento modo in plantam, modo in metallum, silicem, pro seminis directione coagulanda. Quid igitur de quinto & ultimo principio Aqua elementali statuendum erit? atque huic ultro soli, titulum veri principii materialis omnium corporum largimur, atque hoc tanto majori jure ipsi competere arbitramur, quanto certius constat solam aquam elementalem in aliud prius reduci non posse." Ibid, p. 28.

[148] On Jungken, see Partington, ii, pp. 303-4.

[149] Jungken, *Chymia Experimentalis Curiosa. Ex principiis mathematicis demonstrata* (Frankfurt, 1681), pp. 15-18.

restrictions on the use of mechanical explanations. The motion of particles of matter cannot produce the variety of natural bodies, unless they are directed by a spiritual, informative agent, namely the seed.[150] Jungken's explicit eulogies of van Helmont and Boyle do not prevent him from adopting, though in a revised version, the five principles. To him, the chemical principles are not the ultimate constituents of bodies, but are profitably employed in chemistry. The principles differ according to the shape, weight and motions of their particles.[151]

According to Johannes Kunckel, chemist to the Elector of Saxony, water and salt are the ultimate ingredients of all natural bodies, from which mercury, sulphur and earth derive.[152] As far as the chemical non-entities are concerned, Kunckel includes among them the sulphurs of metals and the mercury of vegetables and of animals. He also rejected the universal solvent as a chemical non-entity.[153]

The combination of chemistry and corpuscular theories (as well as the joint influence of Boyle and van Helmont), is central to Jacob Barner's popular textbook, *Chymia Philosophica*. Barner rejected the chemical principles on the grounds that they are produced, not extracted, by fire and can be analysed into water, which is the material principles of bodies, and seed, the formal one. In addition, Barner claimed, the *tria prima* cannot be conceived as principles, as they can be transformed one into the other. Barner also rejected the acid/alkali theory by including a third kind of salt, the neutral salt.[154]

A prolific writer, Georg Wolfgang Wedel combined the doctrine of chemical principles with the corpuscular theory of matter.[155] Wedel's

[150] Jungken, *Chymia experimentalis, sive Medicus praesenti seculo...* (Frankfurt, 1682), p. 3.

[151] Ibid., pp. 7-15.

[152] On Johannes Kunckel (1630-1703), who is known mostly for his works on glass and phosphorus, see Partington, ii, pp. 361-77. Kunckel dealt with the principles of bodies in his *Chymische Anmerckungen* (Wittenberg, 1677), translated into Latin as *Philosophia Chemica* (Amsterdam, 1694). See the English translation, *An Experimental Confirmation of Chymical Philosophy* (London, 1705), p. 14.

[153] Kunckel, *An Experimental Confirmation* (n. 152), p. 162; 122.

[154] J. Barner, *Chymia Philosophica Perfecte Delineata Docte Enucleata & Feliciter Demonstrata* (Nuremberg, 1689), pp. 16-8; 29. Willis's theory is explicitly rejected by Barner (p. 16). On Barner (1641-86), Professor at Leipzig, see Partington, ii, pp. 377-8.

[155] On George Wolfgang Wedel (1645-1721), physician and antiquarian, see DSB; Partington, ii, 315-7; and Hufbauer (n. 115), pp. 165-6.

Experimentum Chimicum Novum (1675) and *Specimen Experimenti Chimici* (1682) deal with the volatile salts of plants – a topic much discussed in the second half of the seventeenth century, witness Coxe's papers published in the *Philosophical Transactions* and the discussions held at the *Académie Royale des Sciences*. Both Wedel's works address the question of the principles of bodies. Like Willis in his *De Fermentantione*, Wedel interpreted the chemical principles in corpuscular terms: spirits are composed of volatile, subtle and active particles, oils of two kinds of particles, i.e. spherical and branched, volatile salts are made of porous corpuscles.[156] Wedel placed special emphasis on the importance of pores in the study of salts. Pores explain the fact that salts easily join the other principles.[157] When dealing with the status of the chemical principles, Wedel follows a *via media*, that is he states that mixed bodies are composed by the chemical principles, not *actu* but *in potentia*, as they are actualised by fire.

GERMAN CORPUSCULARIANS

The use of corpuscular theories in chemistry was particularly relevant in the works of several chemists active in Germany in the second half of the seventeenth century. In the present section I take into account three chemists whose works clearly show the influence of Boyle's ideas: Johann Bohn (1640-1718), Johannes Andrea Stisser (1657-1700) and Friedrich Hoffmann (1660-1742).

In his work on air Bohn adopts some of Boyle's views, notably the latter's explanation of the spring of air.[158] Bohn believes air to be a mixture containing a variety of corpuscles of different nature in continuous movement. *Semina* – active corpuscles – and saline particles abound in air. What makes air a powerful solvent is not, he thinks, some kind of universal spirit, nor the three chemical principles, but the continuous agitation and particular texture of its corpuscles.[159] In *Epistola ad Langellottum* (1675) Bohn quotes Boyle's arguments against the acid/alkali doctrine. He maintains that the standard definitions of these two principles are not

[156] Wedel, *Specimen experimenti chimici novi* (Jena, 1682), pp. 8-12 and 58-9.

[157] Ibid., pp. 60-5.

[158] Bohn, *Meditationes Physico-chymicae De Aeris in Sublunaria Influxu, Scilicet neque secundum Peripatheticos, nec Chymicos* (Leipzig, 1678, 1st edn: 1675), p. 104. On Johann Bohn (1640-1718), Professor of Anatomy in Leipzig, see Partington, ii, pp. 300-2 and Hufbauer (n. 115), p. 165.

[159] Ibid., pp. 56-79; 104.

adequate, as they are circular. The ebullition and effervescence which is considered the signs of an acid-alkali reaction may also be produced by the combination of two acids. The effervescence is in fact the result of the very rapid motions of the corpuscles.[160] The structure of matter is thoroughly discussed in some dissertations held at the University of Leipzig under Bohn's direction. The first dissertation, bearing the title 'De corporum dissolutione' (*respondens* W. Pauli), deals with the origin of fluidity. The explanation of fluidity is based on the motion and texture of corpuscles. Bodies become fluid when there is a discontinuity in their particles, so that fire or air can penetrate the pores. The texture of solid bodies is very close and there are few or no *vacuola* among them. The supposed three chemical principles are not simple and elementary substances. They ultimately consist of invisible corpuscles which possess mechanical properties. Following Boyle, Bohn introduces a classification of corpuscles which justifies his recourse to compound corpuscles (i.e. *prima mixta*) with chemical, not just mechanical properties.[161] This classification of corpuscles may also be found in a second dissertation, entitled 'De corporum combinatione seu concretione'. Chemical analysis does not always yield the ultimate simple corpuscles, but sometimes it separates bodies into *minima secunda* or chemical particles made of different kinds of atoms.[162]

In Stisser's *Acta Laboratorii Chemici* Boyle's corpuscular theory is used to account for the transmutation of metals. By referring to Boyle's *Origine of Formes and Qualities*, Stisser maintains that the transmutation is possible because all bodies are made of one universal matter. Accordingly, a change in the texture can transmute base metals into gold. Though it is more difficult, the degradation of gold can be obtained by altering its texture.[163]

A pupil of Wedel in Jena, Friedrich Hoffmann met Boyle in London in 1684 and became Fellow of the Royal Society in 1719. Hoffmann adopted the corpuscular philosophy and criticised the doctrine of the chemical

[160] Bohn, *Epistola ad Virum Nobilissimum atque Amplissimum D. Joelem Langellottum, Serenissimi Ducis Holstatiae Archiatrum, De Alcali et Acidi insufficientia pro Principiorum seu Elementorum Corporum naturalium* (Leipzig, 1675).

[161] Bohn (*praeses*), *Dissertationes Chymico-Physicae* (Leipzig, 1685): *Dissertatio prima* 'De corporum dissolutione', sig. A1^{v}-B2^{v}.

[162] Ibid., sig. d4^{r-v}.

[163] J. Stisser, *Acta Laboratorii Chemici Specimen Primum* (Helmstaedt, 1690), sig. a1^{v}-c3^{r}. On Johann Andreas Stisser (1657-1700), MD Leiden and Professor of Medicine and Chemistry at Helmstaedt, see Hirsch. (n. 109)

principles. However, he believed that mercury could be extracted from metals.[164] Hoffmann accepted Boyle's analysis of blood as presented in the *Natural History of Blood* (1684). Blood is composed of different kinds of corpuscles: phlegm, spirit, volatile salt, *caput mortuum* and oil. Like Boyle, he did not consider them as the ultimate components of bodies, but as distinct chemical substances found in blood. Hoffmann also hypothesised about the shapes of blood corpuscles, which he supposed were made of branched and globular particles.[165] In medicine, Hoffmann accepted the acid-alkali theory in a weak form. He shared the view that acid and alkali are responsible for a number of diseases, but he also pointed out that they cannot explain every kind of disease; salt is responsible for several diseases, like smallpox and measles.[166]

The mechanical philosophy plays a relevant part in the chemical work of the Swiss physician and chemist Emanuel König (1658-1731), Professor of Physics and then of Medicine at Basle and Fellow of the *Academia Naturae Curiosorum*. The adoption of the mechanical theory of matter does not entail the rejection of the chemical principles. König interprets the principles in terms of corpuscles endowed with distinctive shapes. Salts are made of corpuscles like pointed needles; hence they can easily penetrate other bodies and produce colours, odours and taste in bodies.[167] He distinguished three kinds of salts: acid, alkaline and volatile.[168] König rejected van Helmont's and Boyle's views that plants consist only of water. Water, in König's views is reach in saline and nitrous particles.[169]

[164] On Friedrich Hoffmann (1660-1742) see Partington, ii, pp. 691-700; and Hufbauer (n. 115), pp. 168-9. Hoffmann's views of mercury are in his *Exercitatio Medico-Chymica de Cinnabri Antimonii* (Frankfurt am Main, 1689).

[165] Hoffmann, *Exercitatio acroamatica de acidi et viscidi..* (Frankfurt am Main, 1689), pp. 1-33.

[166] Ibid., pp. 1, 30-33. "Variolas et morbillos inter morbos salinos acres commode referre posse", p. 30.

[167] Joel Langelot and Emanuel König, *Chymia Physica circa corporum Naturalem & Artificialem Statum*, in *Keras Amaltheias, seu Thesaurus Remediorum e triplici Regno*, ed. Emanuele König (Basle, 1693), p. 239. Sulphur is made of large and branched particles, ibid., p. 241.

[168] Ibid., p. 263.

[169] König, *Regnum Vegetabile Physice, Medice, Anatomice, Chymice Theoretice, Practice enucleatum...* (Basle, 1688), p. 12. Evidently, König's objection to Boyle is based on a passage of the *Sceptical Chymist*. As we have seen in chapter four, Boyle did not accept van Helmont's idea that water is a simple substance – though he was positive about water-colture.

ATOMISM AND CHEMISTRY IN ITALY

With the exception of Giordano Bruno, Galileo Galilei and his followers, Italian atomism has received comparatively little attention. Even less attention has been paid to the relationship between atomism and chemistry.[170]

Atomism was promoted by a small number of Renaissance philosophers and physicians, such as Girolamo Fracastoro and Scipione Capece; while, as we have seen, the works of Nifo and Scaliger are important steps towards a corpuscular interpretation of the *minima naturalia*.[171] In the last years of the sixteenth century Paolo Sarpi criticised the Aristotelian doctrine of elements and adopted a corpuscular theory of matter.[172] In addition, the Aristotelian Fortunio Liceti, Professor of Medicine at Padua, explained spontaneous generation in corpuscular terms.[173]

It is apparent that in the first half of the seventeenth century corpuscular theories of matter were often combined with Aristotelian doctrines, and various qualitative versions of atomism flourished throughout the century.[174]

[170] On Bruno's atomism, see P-H. Michel, 'L'atomisme de Giordano Bruno', in La science au seizième siècle (Paris, 1957), pp. 251-63; and H. Gatti,'Notes on Bruno's atomis', in C. Lüthy, J. Murdoch and W.R. Newman (eds.), *Late Medieval and Early Modern Corpuscular Matter Theory*, forthcoming. For Galilei, see W.R. Shea, 'Galileo's Atomic Hypothesis', *Ambix* 17 (1970), 13-27; U. Baldini, 'La struttura della materia nel pensiero di Galilei', *De Homine* 57 (1976), 91-164; P. Redondi, *Galileo: Heretic* (Princeton, 1987). On the Galilean school, see M. Bucciantini and M. Torrini (eds.), *Geometria e Atomismo nella Scuola Galileiana* (Florence, 1992); U. Baldini, 'Il Corpuscolarismo italiano del seicento. Problemi di metodo e prospettive di ricerca', in *Ricerche sull'atomismo del seicento*, (Florence, 1977), pp. 1-76. For a survey of Italian atomism see E. Garin, *Dal Rinascimento all'Illuminismo. Studi e Ricerche* (Pisa, 1970), pp. 79-117.

[171] On Fracastoro, see above, p. 10. For Capece, see DBI.

[172] See P. Sarpi, *Pensieri Naturali, Metafisici e Matematici* (Milano-Napoli, 1996). A reference to Sarpi's atomism is in Campanella's letter to Peiresc of 19 June 1636, published in G. Ernst and E. Canone, 'Una lettera ritrovata: Campanella a Peiresc', *Rivista di storia della filosofia*, 49/2 (1994), 353-66.

[173] F. Liceti, *De Spontaneo Viventium Ortu* (Vicenza, 1618). On Liceti, see J. Roger, *Les sciences de la vie dans la pensée française au XVIIIe siècle* (Paris, 1993^2), pp. 125-7.

[174] Though Galileo's theory of matter was mechanical, he adopted the notion of igneous particles (*ignicoli*). See *Il Saggiatore* (1623), repr. in *Le Opere di Galileo Galilei*, ed. A. Favaro, 20 vols. (Florence, 1890-1909), vi, pp. 351-2. On Galileo's

Though Cartesianism had a significant impact in Italy, one cannot actually find a Cartesian movement there. With a few exceptions, original contributions to matter theory were fairly uncommon and eclecticism prevailed among Italian corpuscular philosophers. Descartes's theories were often combined with Gassendi's, and in the last part of the century, Boyle's corpuscular theories were adopted, especially by members of the *Accademia degli Investiganti*.

Two Frenchmen, Claude Berigard and Johann Chrysostom Magnenus, both of them teaching in Italy, promoted atomism in the first half of the seventeenth century. Berigard's and Magnenus's version of atomism was a qualitative one. They did not share the mechanical outlook and still accepted many aspects of the Aristotelian philosophy. In *Circuli Pisani* (1643) Berigardus adopted the particulate theory of matter without rejecting the Aristotelian notion of form. His corpuscles are all spherical and have different sizes. Berigard's atoms have much in common with Anassagoras's homoeomeries, being corpuscles endowed with specific qualities. According to Berigard, there are infinite species of atoms. All atoms are endowed with an attractive force, analogous to magnetism.[175] They are in mutual contact and there is no vacuum in bodies. For Berigard, all natural bodies (including metals) are generated from living *semina*.[176] Seeds (i.e. clusters of corpuscles) are endowed with attractive force, which he calls sympathy. Teleology is not banished by Berigard, who refers to an external principle (God) directing the motions of corpuscles.[177]

Like Sennert, Magnenus (*Democritus Reviviscens,* 1646) combines atomism with a revised version of the Aristotelian doctrine of elements. However, the difference with Sennert is neat. Whereas chemical experiments play a substantial part in Sennert's atomism, Magnenus makes

ignicoli see E.J. Dijksterhuis, *The Mechanization of the World Picture* (Oxford, 1961), p. 424.

175 C. Berigard, *Circuli Pisani* (Padua, 1661, 1st edn: Udine 1643): 'De Atomis Democriti', p. 61. Berigardus was Professor at the Universities of Pisa and Padua, see DBI, Lasswitz, *Geschichte*, i, pp. 487-98;; N. Badaloni, 'Intorno alla filosofia di Alessandro Marchetti', *Belfagor* 23/5 (1968), 283-316; M. Bellucci, 'La filosofia naturale di Claude Bérigard', *Rivista critica di storia della filosofia* 27 (1971), 363-411.

176 Berigardus, *Circuli Pisani.* (n. 175), p. 131. On Berigard's notion of *semina* see Gemelli, *Aspetti dell'atomismo classico nella filosofia di Francis Bacon e nel Seicento* (Florence, 1996), p. 330.

177 Berigardus, *Circuli Pisani* (n. 175), p. 4.

little use of experiments and does not have recourse to chemistry to confirm his theory of matter. His arguments are almost entirely speculative. Having rejected the Aristotelian view that prime matter exists *in potentia*, Magnenus keeps the notion of form and the doctrine of elements. He reduces the elements to three, fire, water and earth. Air, in his opinion, does not enter into the compounds. For Magnenus, elements have three properties: first, the material bulk (*moles substantialis corporea*); second, their respective primary qualities; third, an appetite, i.e, desire of the *mixtio*.[178] The elements are made of atoms, which Magnenus considers to be simple, homogeneous and indivisible material substances. They have a variety of shapes, which – with the exception of earth (cube) – are not stable. Magenus's atoms have a specific set of qualities which differentiate the atoms of each element: fire atoms are hot and luminous, those of water are moist and transparent, while earth atoms are cold and opaque.

It is apparent that Magnenus's atoms have much in common with the *minima naturalia,* as he conceived atoms as the smallest parts of bodies. Hence there are atoms of the elements, as well as atoms of mixed bodies, including living substances (*atomi viventium*).[179] Atoms are not perfectly solid bodies, as their shapes can be changed, and accordingly they can occupy more or less space. This feature of atoms explains rarefaction and condensation without having recourse to void.[180]

It is clear from Galileo and from two minor figures like Berigard and Magnenus, that investigations on the structure of matter carried out in the first half of the seventeenth century had little or no recourse to chemistry – which indeed played a prominent part in Germany, France and England. The reasons for this difference from other scientific communities can be found

[178] Johann Chrysostomus Magnenus, *Democritus Reviviscens, sive dc Atomis* (The Hague, 1658; first edn: Pavia, 1646) p. 91: "In quovis elemento tria sunt primario consideranda. primo, moles substantialis corporea; secundo, qualitas prima; tertio, appetentia mixtionis." On Magnenus, see Lasswitz, *Geschichte*, i, pp. 498-512 and Gemelli, *Aspetti* (n. 176), pp. 326-9.

[179] "Atomus ignea est entitas corporea, substantialis simplex, & pure homogenea, indivisibilis ex natura sua, calida, & lucida secundum quid, ordinata a natura ad mixtum cum aliis elementorum atomis faciendum: dixi lucida, secundum quid, nam atomi ignae non lucent, nisi certis conditionibus [...] Atomus aquea [...] humida & diaphana. Atomus terrea, frigida & adiaphana." Magnenus, *Democritus* (n. 178), pp. 190-1.

[180] See Gemelli, *Aspetti* (n. 176), pp. 327-9.

both in the status of chemistry in Italy and in the fact that the disciples of Galilei followed their master's distrust of chemistry and alchemy.

In the second half of the sixteenth century Paracelsianism had a limited diffusion and alchemy was dominated by traditional themes – witness Giovanni Agostino Panteo, Giovanni Aurelio Augurelli, Giovanni Bracesco, Domenico Pizzimenti, Giovan Battista Nazari.[181] One can single out only a few naturalists and physicians who adopted Paracelsus's views. Among them Leonardo Fioravanti and Giovan Battista Della Porta were the most influential, while the Veronese physician Zeffirele Tommaso Bovio stands out for his wholehearted adherence to Paracelsianism and for his attacks on Galenic medicine.[182] At the beginning of the seventeenth century some interest in Paracelsianism emerged. Paracelsian doctrines were adopted by members of the Roman *Accademia dei Lincei*, and became object of interest at the Medici court.[183] The relationship of Campanella to Paracelsianism is complex. He eulogised Paracelsus's medicine, but adopted neither his theory of principles, nor his remedies. On the whole, besides some interest in Paracelsus, Campanella never accepted the Paracelsian cosmology.[184]

[181] Cf. G. Carbonelli, *Sulle fonti storiche della chimica e dell'alchimia in Italia* (Rome, 1925) and A. Perifano, *L'Alchimie à la Cour de Côme Ier de Médicis: savoirs, culture et politique* (Paris, 1997).

[182] See Leonardo Fioravanti, *Compendio dei secreti rationali* (Venice, 1564). On Fioravanti see W. Eamon, *Science and the Secrets of Nature* (Princeton, 1994), pp. 182-192 and *passim*; Z.T. Bovio, *Melampigo, ovvero confusione de' medici sofisti* (Verona, 1585). On Bovio see DBI. In his *Magia Naturalis* (1558, 2nd revised edn 1589) Della Porta adopted a number of Paracelsian remedies, besides the pseudo-Paracelsian recipe for the *unguentum armarium*. For Della Porta, see DBI; Partington, ii, pp. 15-25; and W. Eamon, *Science and the Secrets of Nature*, pp. 210-7.

[183] See A. Clericuzio and S. De Renzi, 'Medicine, Alchemy and Natural Philosophy in the Early Accademia dei Lincei', in D.S. Chambers and F. Quiviger, *Italian Academies in the Sixteenth Century* (London, 1995), pp. 175-194 and P. Galluzzi, 'Motivi Paracelsiani nella Toscana di Cosimo II e Don Antonio dei Medici: Alchimia, Medicina, Chimica e Riforma del Sapere', in *Scienze, Credenze Occulte, Livelli di Cultura* (Florence, 1982), pp. 31-62.

[184] See M. Mönnich, *Tommaso Campanella. Sein Beitrag zur Medizin und Pharmazie der Renaissance* (Stuttgart, 1990), pp. 86-89 and M-P. Lerner, 'Campanella et Paracelse', in J-C. Margolin and S. Matton, *Alchimie et Philosophie à la Renaissance* (Paris, 1993), pp. 379-393. Campanella also rejected atomism as, in his view, it was based on matter only, see Campanella to Peiresc, 22 August

In the mid-century appeared a generation of physicians that developed interest in iatrochemistry and in Helmontianism. The key-figure of this movement, which spread in Southern Italy, was Marco Aurelio Severino (1580-1656). His medical works contain evidence of his acceptance of relevant aspects of Paracelsianism, witness his use of chemistry in medical research. Though he did not develop an articulate theory of matter, he adopted some corpuscular views in the study of living organisms. His works also betray a vitalistic conception of nature, which he evidently borrowed from Campanella.[185]

Chemical views may be found in Giovanni Alfonso Borelli's first work, *Delle Cagioni de le febri maligne* (1649), where the origin of epidemics is explained by means of corrosive atoms coming from the bowels of the earth into air.[186] In the same work Borelli held an atomistic theory of contagion, based on Lucretius and Fracastoro.[187] Borelli's theory of matter can be defined as a 'moderate' mechanism. In his works published after 1649 he adhered to the principles of the mechanical philosophy (shape, size and motions of corpuscles account for all phenomena). However, in *De vi percussionis* (1667) Borelli referred to corpuscles of the four elements, as well as to corpuscles of spirit. Borelli was reluctant to formulate philosophical doctrines about the structure of matter. All the same, he

1635, published in T. Campanella, *Lettere*, ed. V. Spampanato (Bari, 1927), p. 322. See also G. Ernst and E. Canone (n. 172).

[185] Severino rejected Epicurus's philosophy as impious. See M.A. Severino, *Zootomia Democritea*, (Nuremberg, 1645), pp. 4-5 and 42. He assumed that all natural bodies are animated (id., p. 34). See also *Antiperipatias. Hoc est adversus Aristoteleos de respiratione piscium diatriba* (Naples, 1659). On Marco Aurelio Severino see DSB; C. Schmitt and C. Webster, 'Marco Aurelio Severino and his relationship to Harvey: some preliminary considerations' in A.G. Debus (ed.) *Science, Medicine and Society in the Renaissance. Essays to honour Walter Pagel*, 2 vols. (London, 1972), pp. 63-72 and id., 'Harvey and M. Aurelio Severino: A Neglected Medical Relationship', *Bulletin of the History of Medicine* 45 (1971), 49-75. Severinus saw Democritus more as a champion of anatomy, than an atomist.

[186] See G.A. Borelli, *Delle Cagioni delle Febbri Maligne della Sicilia negli Anni 1647 e 1648* (Cosenza, 1649), pp. 108-28. For the reference to chemical remedies, see pp. 140-1. Borelli explains digestion as the outcome of a chemical process produced by a solvent. On Borelli, see DSB; DBI and D. Bertoloni-Meli, 'The Neoterics and Political Power in Spanish Italy: Giovanni Alfonso Borelli and his Circle', *History of Science* 34 (1996), 57-89.

[187] Cf. P. Galluzzi, 'G.A. Borelli dal Cimento agli Investiganti', in F. Lomonaco and M. Torrini (eds.), *Galileo e Napoli* (Naples, 1987), pp. 339-55.

stressed that when he had recourse to corpuscles, he referred to compound corpuscles, not the ultimate constituents of bodies. Like Boyle, he criticised van Helmont's water theory, by maintaining that water was no simple substance but a compound one, made of different kinds of corpuscles.[188] In *De motionibus naturalibus* (1667) Borelli presented an elaborate mechanical corpuscular theory of matter, where he explained the spring of air as produced by particles similar to small machines – particles which could be compressed by force.[189] The explanation was mechanical, and Borelli flatly rejected attraction and all actions at a distance. Nevertheless, Borelli's corpuscles were not inert, but were endowed with an internal principle of movement. In his work devoted to physiology (*De Motu Animalium)* mechanical views play a major part. However, he did not rule out iatrochemistry, as may be seen from his explanation of muscular motion in terms of a chemical process, namely fermentation.[190] Borelli's teaching and his works gave a strong impulse to the development of corpuscular philosophy in the last decades of the seventeenth century.

The activity of matter also provided the foundation to the physical theories of Donato Rossetti, a pupil of Borelli at Pisa. In Rossetti's theory of matter the mechanical affections of particles (shape and size) play a marginal part; the central role is played by the notion of force. He distinguished two kinds of atoms – dark and bright. All atoms are endowed with activity and forces (both attractive and repulsive) and form molecules. Like magnets, all atoms have two poles and their energy is concentrated in the centre of the corpuscles, while on the external surface it is weaker and decreases with the distance from the centre. Molecules too are endowed with forces and attract one another. The stronger their energy, the harder is the body they form.[191] Different degrees of energy account for the various

[188] G.A. Borelli, *De Vi Percussionis* (Bologna, 1667), pp. 189 and 236-7, where he speaks of "ignis particulae minimae". He also refers to "corpuscula magnetica" and to "corpuscula spirituosa", ibid., p. 242. Borelli's attack on van Helmont occurs in *De Motu Animalium* (Rome, 1680), p. 179. The best study of Borelli's theory of matter is still Lasswitz, *Geschichte*, ii, pp. 300-28.

[189] G.A. Borelli, *De motionibus naturalibus a gravitate pendentibus* (Bologna, 1670).

[190] Borelli, *De Motu Animalium* (n. 188), pp. 261-4.

[191] D. Rossetti, *Composizione e passione de' vetri* (Livorno, 1671), pp. 1-14. On Donato Rossetti (1633-1688), see S. Gómez López, *Le passioni degli atomi. Montanari e Rossetti. Una polemica tra galileiani* (Florence, 1997), dealing with the discussions between Geminiano Montanari, a strict mechanist, and Rossetti. See

combinations of atoms and molecules.[192] According to Rossetti, energy is diffused in all natural bodies. Following a pattern which was familiar to both Kepler and the Paracelsians, Rossetti assumed that the earth was a living body. Metals are continuously generated in the bowels of the earth. Like an animal's heart, the centre of the earth is the source of heat and of life.[193]

The combination of corpuscular theory and chemistry is to be found mostly among the members of the Neapolitan *Accademia degli Investiganti*.[194] A key figure of southern Italy science, Tommaso Cornelio adopted the corpuscular philosophy and promoted the *Accademia degli Investiganti*. In his early career, as attested by the *De Cognatione Aëris et Aquae* (written in 1649), Cornelio was influenced by Descartes. Cornelio, who maintained that matter and motion were the two principles of natural philosophy, described water particles as longish and smooth (*oblungae, ac teretes*). Like Descartes, Cornelio rejected vacuum, stating that an ethereal substance filled the spaces among larger particles.[195] In *Progymnasmata*, his *opus magnum*, Cornelio's theory of matter became less mechanical. Like Gassendi, Cornelio held the view that matter is active and put some restrictions on the use of mathematical and mechanical theories in the study of nature, notably in physiology. Cornelio explains life by by means of vital spirits, which are described as a volatile salt, and digestion is conceived as a chemical process involving fermentation.[196] Following Glisson, Cornelio maintained that the parts of living organisms and plants are endowed with irritability – originating from spirits. Spirits are the source of energy to be found in matter. Whereas Cornelio had no interest in Helmontian doctrines, the other *Investiganti* often had recourse to van Helmont in their works.

Unlike Paracelsianism, the doctrines of van Helmont were widely disseminated in Italy. Helmontianism was promoted by the German

Montanari, 'Della Natura, et Uso degli Atomi', in S. Rotta, 'Scienza e pubblica felicità in Geminiano Montanari', *Miscellanea Seicento* 2/2 (1971), 187-195.

[192] Rossetti, *Composizione e passione* (n. 191), p. 16.

[193] Rossetti, *Antignome fisico-matematiche* (Livorno, 1667). See Gómez López, *Le passioni* (n. 191), pp. 97-101.

[194] See M.H. Fish, 'The Academy of Investigators', in E.A. Underwood (ed.), *Science, Medicine and History*, 2 vols. (Oxford, 1953), i, pp. 521-63.

[195] On Tommaso Cornelio (1614-86) see DBI and M. Torrini, *Tommaso Cornelio e la ricostruzione della scienza* (Naples, 1977).

[196] T. Cornelio, *Progymnasmata Physica* (Venice, 1663): 'De ratione philosophandi', pp. 25-27 and 'De vita', pp. 101-2 and 'De nutricatione', pp. 79-84.

physician Otto Tachenius, who settled in Venice where he published his iatrochemical works. Ludovico Conti, physician and alchemist from Macerata, practised in Venice. Conti held Helmontian views and claimed that he had prepared the *Alkahest*. Helmontian medicine was supported in Naples by Sebastiano Bartoli, who vehemently opposed traditional medicine.[197] Iatrochemistry played a central part in the works of Leonardo di Capua, a prominent member of the *Accademia degli Investiganti*. His adherence to iatrochemistry is demonstrated by the *Lezioni intorno alla Natura delle Mofete*, a series of lectures di Capua delivered at the *Accademia degli Investiganti* devoted to the study of poisonous exhalations near Naples. Respiration is the main topic of di Capua's *Lezioni*. Like Boyle, he considers air an elastic and chemically complex substance, containing corpuscles of different nature. Some of them are essential to life which di Capua believes is produced by fermentation, namely a rapid motion of corpuscles of different substances.[198] Di Capua explains both chemical reactions and physiological processes in corpuscular terms. Besides having mechanical affections, di Capua's corpuscles carry distinct chemical properties. An ethereal substance, according to him, keeps all corpuscles in constant movement, and, as a result, effluvia are produced by all natural bodies.[199] Chemical processes are introduced to explain human physiology. As in the alembic, a distillation occurs in the brain, where blood is distilled and a volatile alkaline fluid is generated. This fluid is enclosed in the nerves and is responsible for sensation and movements.[200] In his subsequent work di Capua traces a history of medicine which highlights the role of iatrochemistry, and in particular the roles of van Helmont and Willis. Di Capua's conclusions, however, stress the uncertainty of medicine – a view which is ultimately directed against his contemporary Galenists.[201]

Among the members of the *Accademia degli Investiganti*, Lucantonio Porzio stands out for the variety of his research, which included geometry, physics, chemistry and medicine. He adopted Boyle's theory of the spring of air and pursued experimental investigations on air and respiration. His corpuscular theory of matter is evidently based on Boyle's. In *Erasistratus*,

[197] On Tachenius see Partington, ii, pp. 291-7; on Conti see Ferguson, ii, pp. 173; on Sebastiano Bartoli (1630-1676) see Fisch, 'The Academy' (n. 194), pp. 524-5.

[198] L. di Capua, *Lezioni intorno alla natura delle mofete* (Naples, 1683), pp. 77-80. On di Capua (1617-95), see DBI.

[199] Di Capua, *Lezioni* (n. 198), p. 139

[200] Ibid., pp. 150-5.

[201] Di Capua, *Parere divisato in otto ragionamenti* (Naples, 1681).

which is a dialogue on phlebotomy, Porzio explains physiological processes in chemical terms, but he does not commit himself to the chemical doctrine of principles.[202]

The activities of the Neapolitan group met with strong opposition from both medical and religious sides. Cartesianism and atomism were attacked by the Jesuit Giovan Battista de Benedictis (alias Benedetto Aletino) in *Lettere Apologetiche in difesa della teologia scolastica e della filosofia peripatetica* (1694) as incompatible with the Catholic doctrine of the Eucharist.[203] The outcome was a trial against 'atheists' (1688-1697), involving members of the *Accademia degli Investiganti*.[204]

CONCLUSION

On the basis of the available evidence it is possible to affirm that the non-mechanical atomism did not disappear in the last decades of the seventeenth century. Far from being a backward version of the corpuscular philosophy, it still played an important part in science, especially in chemistry and in medicine.

Boyle's chemical ideas had a strong impact in the last decades of the century. Both in England and on the Continent Boyle's views were often combined with Helmontianism. Boyle's arguments against the Paracelsian principles brought about a reinterpretation of their status. The position of Lemery is representative of a widespread attitude towards chemical principles – especially in France and in the Netherlands. Though Lemery

[202] L. Porzio, *Erasistratus, sive de sanguinis missione* (Rome, 1682). The characters of Porzio's dialogue are Erasistratus, Galen, van Helmont and Willis. Lucantonio Porzio, who was Professor of Medicine in Rome, was also connected with the *Accademia di Medinacoeli* in Naples and with the Academy of Paolo Sarotti in Venice. On Lucantonio Porzio, see M. Torrini, *Dopo Galileo Una polemica scientifica (1684-1711)* (Florence, 1979); A. Dini, *Filosofia della Natura, Medicina, Religione. Lucantonio Porzio (1639-1724)* (Milan, 1985). In Venice Porzio performed Boyle's experiments on air and respiration, see C. Pighetti, *L'influsso scientifico di Robert Boyle nel tardo '600 italiano* (Milan, 1988), pp. 147-54. Pighetti's book does not deal with the influence of Boyle's chemistry.

[203] See P. Rossi, 'I punti di Zenone: una preistoria vichiana', *Nuncius* 13/2 (1998), 377-425. See also Redondi, (n. 170). Atomism was advocated by Francesco d'Andrea, see M. Torrini, 'Atomi in Arcadia', *Nouvelles de la République des Lettres*, 1984, 81-95; A. Borrelli, *D'Andrea atomista. L' "Apologia" e altri inediti nella polemica filosofica della Napoli di fine Seicento* (Naples, 1995).

[204] Cf. Dini, *Filosofia* (n. 202), pp. 100-110.

did not accept Boyle's (and van Helmont's) view that the so-called principles were not extracted, but produced by fire, he recognised that they could not be obtained from all substances. In addition, Lemery did not consider them as the ultimate constituents of bodies. As he put it, they are principles for us, not in nature. Boyle's views were also relevant in the discussions about the acid/alkali theory. As a result of Boyle's criticism (the theory was too comprehensive and the definitions of the two substances were circular), both André and Bertrand modified this theory. It was no longer a general chemical theory and the existence of neutral salts was admitted.

The classification of corpuscles, which played an important part in Boyle's work, was developed by a number of chemists. The most interesting developments of Boyle's classification are in the writings of Joachim Becher and of Daniel Coxe. The latter made consistent use of the notion of molecule – namely a compound corpuscle with chemical properties.

EPILOGUE

From the beginning of the seventeenth century atomism and chemistry were closely linked. Chemistry produced the best experimental evidence to support atomism; and the particulate theory of matter became more and more influential in the development of chemistry. A variety of reactions were explained in terms of addition and subtraction of corpuscles. It has been possible to trace a continual tradition of chemical atomism from Sala and Sennert to Boyle and Lemery.

This does not mean that Boyle's achievement in chemistry is negligible, nor does it deny the originality of his theories. Rather, both are better understood in the context of the chemical atomism. Despite the similarities, the differences between Boyle and his predecessors are relevant. First, Boyle unambiguously rejected – with both theoretical and experimental arguments – the Aristotelian elements, the chemical principles and the substantial forms. Second, he developed and articulated a coherent corpuscular theory of matter which he employed to explain a variety of phenomena, in physics, chemistry and medicine. Analyses of his works, both in print and in manuscript, have shown that Boyle's theory of matter cannot be labelled as strictly mechanical. Though he repeatedly eulogised the mechanical philosophy as the simplest and the most intelligible theory of matter, he employed the seminal principles, which in fact do not form a part of the mechanical philosophers' ontology. Moreover, when he dealt with chemistry and medicine, he refrained from reducing all phenomena to the shape, size and motions of the primary corpuscles. Boyle preferred to refer physical properties of bodies (i.e. the spring of air, fluidity, magnetism, electricity, cohesion) to the texture of corpuscles. Evidently he was aware that the Cartesians' explanations, which were based on specific – and at the same time arbitrary – sizes and shapes of the particles, were based on little or no experimental evidence. Moreover, the reductionism along the lines posited by Descartes failed to account for the puzzling question of powers to be found in bodies. For Boyle, corpuscles can produce a given quality only if they act on bodies which have a texture fit to be acted upon by them. He explained this view by means of the analogy with the key and lock. The size and shape of the key is relevant, but is not adequate to explain its power to open the lock. The power of the key derives from the relationship of its shape and size to the shape and size of the lock. When Boyle dealt with chemistry and medicine, his distance from the strict mechanical philosophy becomes more apparent. Boyle's explanations employ chemical notions and are seldom reduced to the shape and size of corpuscles. He spoke of acid, spirituous, alkaline corpuscles. The classification of corpuscles he established provides the foundation to the use of chemical corpuscles. Once we have reassessed the main features of Boyle's corpuscular philosophy, it has become possible to understand some complex aspects of his chemical work, namely where he stands as regards the chemical principles. He unambiguously rejected the chemists' claim that their principles were simple, homogenous substances which could be extracted from all natural bodies. However, it has become apparent that Boyle's position on the chemical principles is more complex than historians have hitherto thought. Boyle's rejection of the chemists' doctrine of principles did not entail that he ruled out the possibility of discovering simple and homogeneous substances. His rejection of the traditional analysis by fire and his quest for new methods of chemical analysis (including the universal solvent) are the grounds of his view that simple and homogeneous substances could indeed be found. This clarification can make sense of Boyle's somewhat puzzling views of mercury – which he saw as a homogeneous substance contained in metals. The fact that Boyle did not rule out (at least in theory) the existence of simple and homogeneous chemical substance is not incompatible with his corpuscular theory of matter. The classification of corpuscles which has been insisted upon in the present book provides the foundation to his notion of chemically homogeneous substances.

Examination of primary sources has shown that the traditional interpretation of seventeenth-century atomism, as developing from a qualitative version to a purely quantitative and mechanical, is contradicted by strong discrepant evidence. Corpuscular theories can be different from the pure mechanical philosophy. In fact Sennert, Basso, Gassendi and Boyle (to mention only some of the better known figures I have dealt with) can be considered as representative of different versions of corpuscular philosophy, but none can be defined as strictly mechanist. Far from being an obstacle to the development of atomism, the non-mechanical version of the corpuscular theory provided a viable *explanans* of a variety of phenomena, both in chemistry and in medicine.

Since the view that corpuscular and vitalistic are two incompatible theories of matter has long dominated the historians' interpretations of early modern science, a reconsideration of this topic is necessary, in order to avoid frequent oversimplifications. On the basis of the present study, it is possible to suggest a tentative classification of the main theories of matter which prevailed in early modern times. It goes without saying that the following classification does not pretend to account for of all the varieties of view-points held in the period under consideration. In addition, it must be stressed that the obvious fact that the authors included in the same category might have expressed different philosophical conceptions on a number of themes is not relevant for the present classification – which is meant only to provide a conceptual tool to put some order in the *mare magnum* of early modern matter theories. The following categories may be employed to this purpose:

1. Purely vitalistic theories of matter: matter is endowed with sympathies, antipathies, attractions, forces and powers. Corpuscles are not included. These were the views of Telesio, Paracelsus and Campanella.

2 Vitalistic theories which adopt some sort of particulate theory of matter: Bruno, d'Espagnet, J.B. van Helmont.

3. Corpuscular views which employ the notion of form as the principle of organisation of matter: Sennert.

4. Corpuscular theories which include the notion of active principles, which may be divided into two distinct groups differing in their opinions on the origin of motion: 4a. Matter is active and atoms are endowed with motion from the beginning (Gassendi, M. Cavendish, Charleton, Highmore, Willis); 4b. matter is not active, but some corpuscles (i.e. the seminal principles) are endowed by God with plastic power (Boyle).

5. Corpuscular theories which adopt a purely mechanical view of matter: matter is inert and all interactions are produced by the impact of corpuscles having only mechanical properties, that is, size and shape (Descartes, Spinoza, Huygens and Hartsoeker).

In the seventeenth century existed a variety of versions of the corpuscular theory of matter, and the mechanical philosophy was just one of them. Among the philosophers who adopted a particulate theory, some did not share the idea that matter had purely mechanical properties. Many natural philosophers and physicians maintained that corpuscles were endowed with forces and active principles, as well as with chemical properties, which they did not reduce to the mechanical attributes of particles. The non-mechanical atomism, hitherto often-overlooked, was in fact widely adopted by seventeenth-century chemists. The fact that the majority of chemists did not reject the chemical principles does not mean that they ignored Boyle's arguments. Indeed, a closer look at late seventeenth-century chemistry has made it clear that Boyle's experiments and theories have brought about an operational interpretation of the principles. They were not considered as simple substances; but as the product of chemical analysis. In addition, it is apparent that Boyle's works have enhanced the combination of chemistry with corpuscular theories, which was widespread in the last decades of the century.

Finally, it is my hope that the understanding of the interactions between chemistry and corpuscular philosophy, and the reassessment of the role of non-mechanical atomism in the seventeenth century may help provide the background to the study of Newton's chemistry and theory of matter – a subject which is beyond the scope of the present book.

INDEX

ARCHIVES INTERNATIONALES D'HISTOIRE DES IDÉES
*
INTERNATIONAL ARCHIVES OF THE HISTORY OF IDEAS

1. E. Labrousse: *Pierre Bayle.* Tome I: *Du pays de foix à la cité d'Erasme.* 1963; 2nd printing 1984 ISBN 90-247-3136-4
   For Tome II *see below under Volume 6.*
2. P. Merlan: *Monopsychism, Mysticism, Metaconsciousness.* Problems of the Soul in the Neoaristotelian and Neoplatonic Tradition. 1963; 2nd printing 1969 ISBN 90-247-0178-3
3. H.G. van Leeuwen: *The Problem of Certainty in English Thought, 1630–1690.* With a Preface by R.H. Popkin. 1963; 2nd printing 1970 ISBN 90-247-0179-1
4. P.W. Janssen: *Les origines de la réforme des Carmes en France au 17e Siècle.* 1963; 2nd printing 1969 ISBN 90-247-0180-5
5. G. Sebba: *Bibliographia Cartesiana.* A Critical Guide to the Descartes Literature (1800–1960). 1964 ISBN 90-247-0181-3
6. E. Labrousse: *Pierre Bayle.* Tome II: *Heterodoxie et rigorisme.* 1964 ISBN 90-247-0182-1
7. K.W. Swart: *The Sense of Decadence in 19th-Century France.* 1964 ISBN 90-247-0183-X
8. W. Rex: *Essays on Pierre Bayle and Religious Controversy.* 1965 ISBN 90-247-0184-8
9. E. Heier: *L.H. Nicolay (1737–1820) and His Contemporaries.* Diderot, Rousseau, Voltaire, Gluck, Metastasio, Galiani, D'Escherny, Gessner, Bodmer, Lavater, Wieland, Frederick II, Falconet, W. Robertson, Paul I, Cagliostro, Gellert, Winckelmann, Poinsinet, Lloyd, Sanchez, Masson, and Others. 1965 ISBN 90-247-0185-6
10. H.M. Bracken: *The Early Reception of Berkeley's Immaterialism, 1710–1733.* [1958] Rev. ed. 1965 ISBN 90-247-0186-4
11. R.A. Watson: *The Downfall of Cartesianism, 1673–1712.* A Study of Epistemological Issues in Late 17th-Century Cartesianism. 1966 ISBN 90-247-0187-2
12. R. Descartes: *Regulæ ad Directionem Ingenii.* Texte critique établi par Giovanni Crapulli avec la version hollandaise du 17e siècle. 1966 ISBN 90-247-0188-0
13. J. Chapelain: *Soixante-dix-sept Lettres inédites à Nicolas Heinsius (1649–1658).* Publiées d'après le manuscrit de Leyde avec une introduction et des notes par B. Bray. 1966 ISBN 90-247-0189-9
14. C. B. Brush: *Montaigne and Bayle.* Variations on the Theme of Skepticism. 1966 ISBN 90-247-0190-2
15. B. Neveu: *Un historien à l'Ecole de Port-Royal.* Sébastien le Nain de Tillemont (1637–1698). 1966 ISBN 90-247-0191-0
16. A. Faivre: *Kirchberger et l'Illuminisme du 18e siècle.* 1966 ISBN 90-247-0192-9
17. J.A. Clarke: *Huguenot Warrior.* The Life and Times of Henri de Rohan (1579–1638). 1966 ISBN 90-247-0193-7
18. S. Kinser: *The Works of Jacques-Auguste de Thou.* 1966 ISBN 90-247-0194-5
19. E.F. Hirsch: *Damião de Gois.* The Life and Thought of a Portuguese Humanist (1502–1574). 1967 ISBN 90-247-0195-3
20. P.J.S. Whitemore: *The Order of Minims in 17th-Century France.* 1967 ISBN 90-247-0196-1
21. H. Hillenaar: *Fénelon et les Jésuites.* 1967 ISBN 90-247-0197-X
22. W.N. Hargreaves-Mawdsley: *The English Della Cruscans and Their Time, 1783–1828.* 1967 ISBN 90-247-0198-8
23. C.B. Schmitt: *Gianfrancesco Pico della Mirandola (1469–1533) and his Critique of Aristotle.* 1967 ISBN 90-247-0199-6
24. H.B. White: *Peace among the Willows.* The Political Philosophy of Francis Bacon. 1968 ISBN 90-247-0200-3

ARCHIVES INTERNATIONALES D'HISTOIRE DES IDÉES
*
INTERNATIONAL ARCHIVES OF THE HISTORY OF IDEAS

25. L. Apt: *Louis-Philippe de Ségur.* An Intellectual in a Revolutionary Age. 1969 ISBN 90-247-0201-1
26. E.H. Kadler: *Literary Figures in French Drama (1784–1834).* 1969 ISBN 90-247-0202-X
27. G. Postel: *Le Thrésor des prophéties de l'univers.* Manuscrit publié avec une introduction et des notes par F. Secret. 1969 ISBN 90-247-0203-8
28. E.G. Boscherini: *Lexicon Spinozanum.* 2 vols., 1970 Set ISBN 90-247-0205-4
29. C.A. Bolton: *Church Reform in 18th-Century Italy.* The Synod of Pistoia (1786). 1969 ISBN 90-247-0208-9
30. D. Janicaud: *Une généalogie du spiritualisme français.* Aux sources du bergsonisme: [Félix] Ravaisson [1813–1900] et la métaphysique. 1969 ISBN 90-247-0209-7
31. J.-E. d'Angers: *L'Humanisme chrétien au 17ᵉ siècle.* St. François de Sales et Yves de Paris. 1970 ISBN 90-247-0210-0
32. H.B. White: *Copp'd Hills towards Heaven.* Shakespeare and the Classical Polity. 1970 ISBN 90-247-0250-X
33. P.J. Olscamp: *The Moral Philosophy of George Berkeley.* 1970 ISBN 90-247-0303-4
34. C.G. Noreña: *Juan Luis Vives (1492–1540).* 1970 ISBN 90-247-5008-3
35. J. O'Higgens: *Anthony Collins (1676–1729), the Man and His World.* 1970 ISBN 90-247-5007-5
36. F.T. Brechka: *Gerard van Swieten and His World (1700–1772).* 1970 ISBN 90-247-5009-1
37. M.H. Waddicor: *Montesquieu and the Pilosophy of Natural Law.* 1970 ISBN 90-247-5039-3
38. O.R. Bloch: *La Philosophie de Gassendi (1592–1655).* Nominalisme, matérialisme et métaphysique. 1971 ISBN 90-247-5035-0
39. J. Hoyles: *The Waning of the Renaissance (1640–1740).* Studies in the Thought and Poetry of Henry More, John Norris and Isaac Watts. 1971 ISBN 90-247-5077-6
    For Henry More, *see also below under Volume 122 and 127.*
40. H. Bots: *Correspondance de Jacques Dupuy et de Nicolas Heinsius (1646–1656).* 1971 ISBN 90-247-5092-X
41. W.C. Lehmann: *Henry Home, Lord Kames, and the Scottish Enlightenment.* A Study in National Character and in the History of Ideas. 1971 ISBN 90-247-5018-0
42. C. Kramer: *Emmery de Lyere et Marnix de Sainte Aldegonde.* Un admirateur de Sébastien Franck et de Montaigne aux prises avec le champion des calvinistes néerlandais.[Avec le texte d'Emmery de Lyere:] *Antidote ou contrepoison contre les conseils sanguinaires et envinemez de Philippe de Marnix Sr. de Ste. Aldegonde.* 1971 ISBN 90-247-5136-5
43. P. Dibon: *Inventaire de la correspondance (1595–1650) d'André Rivet (1572–1651).* 1971 ISBN 90-247-5112-8
44. K.A. Kottman: *Law and Apocalypse.* The Moral Thought of Luis de Leon (1527?–1591). 1972 ISBN 90-247-1183-5
45. F.G. Nauen: *Revolution, Idealism and Human Freedom.* Schelling, Hölderlin and Hegel, and the Crisis of Early German Idealism. 1971 ISBN 90-247-5117-9
46. H. Jensen: *Motivation and the Moral Sense in Francis Hutcheson's* [1694–1746] *Ethical Theory.* 1971 ISBN 90-247-1187-8
47. A. Rosenberg: *[Simon] Tyssot de Patot and His Work (1655–1738).* 1972 ISBN 90-247-1199-1
48. C. Walton: *De la recherche du bien.* A study of [Nicolas de] Malebranche's [1638–1715] Science of Ethics. 1972 ISBN 90-247-1205-X

ARCHIVES INTERNATIONALES D'HISTOIRE DES IDÉES
*
INTERNATIONAL ARCHIVES OF THE HISTORY OF IDEAS

49. P.J.S. Whitmore (ed.): *A 17th-Century Exposure of Superstition.* Select Text of Claude Pithoys (1587–1676). 1972 ISBN 90-247-1298-X
50. A. Sauvy: *Livres saisis à Paris entre 1678 et 1701.* D'après une étude préliminaire de Motoko Ninomiya. 1972 ISBN 90-247-1347-1
51. W.R. Redmond: *Bibliography of the Philosophy in the Iberian Colonies of America.* 1972 ISBN 90-247-1190-8
52. C.B. Schmitt: *Cicero Scepticus.* A Study of the Influence of the *Academica* in the Renaissance. 1972 ISBN 90-247-1299-8
53. J. Hoyles: *The Edges of Augustanism.* The Aesthetics of Spirituality in Thomas Ken, John Byrom and William Law. 1972 ISBN 90-247-1317-X
54. J. Bruggeman and A.J. van de Ven (éds.): *Inventaire* des pièces d'Archives françaises se rapportant à l'Abbaye de Port-Royal des Champs et son cercle et à la Résistance contre la Bulle *Unigenitus* et à l'Appel. 1972 ISBN 90-247-5122-5
55. J.W. Montgomery: *Cross and Crucible.* Johann Valentin Andreae (1586–1654), Phoenix of the Theologians. Volume I: Andreae's Life, World-View, and Relations with Rosicrucianism and Alchemy; Volume II: The *Chymische Hochzeit* with Notes and Commentary. 1973 Set ISBN 90-247-5054-7
56. O. Lutaud: *Des révolutions d'Angleterre à la Révolution française.* Le tyrannicide & *Killing No Murder* (Cromwell, *Athalie*, Bonaparte). 1973 ISBN 90-247-1509-1
57. F. Duchesneau: *L'Empirisme de Locke.* 1973 ISBN 90-247-1349-8
58. R. Simon (éd.): *Henry de Boulainviller* – Œuvres Philosophiques, Tome I. 1973 ISBN 90-247-1332-3

    For Œuvres Philosophiques, Tome II *see below under Volume 70.*
59. E.E. Harris: *Salvation from Despair.* A Reappraisal of Spinoza's Philosophy. 1973 ISBN 90-247-5158-6
60. J.-F. Battail: *L'Avocat philosophe Géraud de Cordemoy (1626–1684).* 1973 ISBN 90-247-1542-3
61. T. Liu: *Discord in Zion.* The Puritan Divines and the Puritan Revolution (1640–1660). 1973 ISBN 90-247-5156-X
62. A. Strugnell: *Diderot's Politics.* A Study of the Evolution of Diderot's Political Thought after the *Encyclopédie.* 1973 ISBN 90-247-1540-7
63. G. Defaux: *Pantagruel et les Sophistes.* Contribution à l'histoire de l'humanisme chrétien au 16e siècle. 1973 ISBN 90-247-1566-0
64. G. Planty-Bonjour: *Hegel et la pensée philosophique en Russie (1830–1917).* 1974 ISBN 90-247-1576-8
65. R.J. Brook: *[George] Berkeley's Philosophy of Science.* 1973 ISBN 90-247-1555-5
66. T.E. Jessop: *A Bibliography of George Berkeley.* With: *Inventory of Berkeley's Manuscript Remains* by A.A. Luce. 2nd revised and enlarged ed. 1973 ISBN 90-247-1577-6
67. E.I. Perry: *From Theology to History.* French Religious Controversy and the Revocation of the Edict of Nantes. 1973 ISBN 90-247-1578-4
68. P. Dibbon, H. Bots et E. Bots-Estourgie: *Inventaire de la correspondance (1631–1671) de Johannes Fredericus Gronovius* [1611–1671]. 1974 ISBN 90-247-1600-4
69. A.B. Collins: *The Secular is Sacred.* Platonism and Thomism in Marsilio Ficino's *Platonic Theology.* 1974 ISBN 90-247-1588-1

ARCHIVES INTERNATIONALES D'HISTOIRE DES IDÉES
*
INTERNATIONAL ARCHIVES OF THE HISTORY OF IDEAS

70. R. Simon (éd.): *Henry de Boulainviller.* Œuvres Philosophiques, Tome II. 1975
ISBN 90-247-1633-0
For Œuvres Philosophiques, Tome I *see under Volume 58.*
71. J.A.G. Tans et H. Schmitz du Moulin: *Pasquier Quesnel devant la Congrégation de l'Index.* Correspondance avec Francesco Barberini et mémoires sur la mise à l'Index de son édition des Œuvres de Saint Léon, publiés avec introduction et annotations. 1974 ISBN 90-247-1661-6
72. J.W. Carven: *Napoleon and the Lazarists (1804–1809).* 1974 ISBN 90-247-1667-5
73. G. Symcox: *The Crisis of French Sea Power (1688–1697).* From the *Guerre d'Escadre* to the *Guerre de Course.* 1974 ISBN 90-247-1645-4
74. R. MacGillivray: *Restoration Historians and the English Civil War.* 1974
ISBN 90-247-1678-0
75. A. Soman (ed.): *The Massacre of St. Bartholomew.* Reappraisals and Documents. 1974
ISBN 90-247-1652-7
76. R.E. Wanner: *Claude Fleury (1640–1723) as an Educational Historiographer and Thinker.* With an Introduction by W.W. Brickman. 1975 ISBN 90-247-1684-5
77. R.T. Carroll: *The Common-Sense Philosophy of Religion of Bishop Edward Stillingfleet (1635–1699).* 1975 ISBN 90-247-1647-0
78. J. Macary: *Masque et lumières au 18ᵉ [siècle].* André-François Deslandes, Citoyen et philosophe (1689–1757). 1975 ISBN 90-247-1698-5
79. S.M. Mason: *Montesquieu's Idea of Justice.* 1975 ISBN 90-247-1670-5
80. D.J.H. van Elden: *Esprits fins et esprits géométriques dans les portraits de Saint-Simon.* Contributions à l'étude du vocabulaire et du style. 1975 ISBN 90-247-1726-4
81. I. Primer (ed.): *Mandeville Studies.* New Explorations in the Art and Thought of Dr Bernard Mandeville (1670–1733). 1975 ISBN 90-247-1686-1
82. C.G. Noreña: *Studies in Spanish Renaissance Thought.* 1975 ISBN 90-247-1727-2
83. G. Wilson: *A Medievalist in the 18th Century.* Le Grand d'Aussy and the Fabliaux ou Contes. 1975 ISBN 90-247-1782-5
84. J.-R. Armogathe: *Theologia Cartesiana.* L'explication physique de l'Eucharistie chez Descartes et Dom Robert Desgabets. 1977 ISBN 90-247-1869-4
85. Bérault Stuart, Seigneur d'Aubigny: *Traité sur l'art de la guerre.* Introduction et édition par Élie de Comminges. 1976 ISBN 90-247-1871-6
86. S.L. Kaplan: *Bread, Politics and Political Economy in the Reign of Louis XV.* 2 vols., 1976
Set ISBN 90-247-1873-2
87. M. Lienhard (ed.): *The Origins and Characteristics of Anabaptism / Les débuts et les caractéristiques de l'Anabaptisme.* With an Extensive Bibliography / Avec une bibliographie détaillée. 1977 ISBN 90-247-1896-1
88. R. Descartes: *Règles utiles et claires pour la direction de l'esprit en la recherche de la vérité.* Traduction selon le lexique cartésien, et annotation conceptuelle par J.-L. Marion. Avec des notes mathématiques de P. Costabel. 1977 ISBN 90-247-1907-0
89. K. Hardesty: *The 'Supplément' to the 'Encyclopédie'.* [Diderot et d'Alembert]. 1977
ISBN 90-247-1965-8
90. H.B. White: *Antiquity Forgot.* Essays on Shakespeare, [Francis] Bacon, and Rembrandt. 1978
ISBN 90-247-1971-2
91. P.B.M. Blaas: *Continuity and Anachronism.* Parliamentary and Constitutional Development in Whig Historiography and in the Anti-Whig Reaction between 1890 and 1930. 1978
ISBN 90-247-2063-X

ARCHIVES INTERNATIONALES D'HISTOIRE DES IDÉES
*
INTERNATIONAL ARCHIVES OF THE HISTORY OF IDEAS

92. S.L. Kaplan (ed.): *La Bagarre.* Ferdinando Galiani's (1728–1787) 'Lost' Parody. With an Introduction by the Editor. 1979 ISBN 90-247-2125-3
93. E. McNiven Hine: *A Critical Study of [Étienne Bonnot de] Condillac's* [1714–1780]*'Traité des Systèmes'.* 1979 ISBN 90-247-2120-2
94. M.R.G. Spiller: *Concerning Natural Experimental Philosphy.* Meric Casaubon [1599–1671] and the Royal Society. 1980 ISBN 90-247-2414-7
95. F. Duchesneau: *La physiologie des Lumières.* Empirisme, modèles et théories. 1982 ISBN 90-247-2500-3
96. M. Heyd: *Between Orthodoxy and the Enlightenment.* Jean-Robert Chouet [1642–1731] and the Introduction of Cartesian Science in the Academy of Geneva. 1982 ISBN 90-247-2508-9
97. James O'Higgins: *Yves de Vallone* [1666/7–1705]: *The Making of an Esprit Fort.* 1982 ISBN 90-247-2520-8
98. M.L. Kuntz: *Guillaume Postel* [1510–1581]. Prophet of the Restitution of All Things. His Life and Thought. 1981 ISBN 90-247-2523-2
99. A. Rosenberg: *Nicolas Gueudeville and His Work (1652–172?).* 1982 ISBN 90-247-2533-X
100. S.L. Jaki: *Uneasy Genius: The Life and Work of Pierre Duhem* [1861-1916]. 1984 ISBN 90-247-2897-5; Pb (1987) 90-247-3532-7
101. Anne Conway [1631–1679]: *The Principles of the Most Ancient Modern Philosophy.* Edited and with an Introduction by P. Loptson. 1982 ISBN 90-247-2671-9
102. E.C. Patterson: *[Mrs.] Mary [Fairfax Greig] Sommerville* [1780–1872] *and the Cultivation of Science (1815–1840).* 1983 ISBN 90-247-2823-1
103. C.J. Berry: *Hume, Hegel and Human Nature.* 1982 ISBN 90-247-2682-4
104. C.J. Betts: *Early Deism in France.* From the so-called 'déistes' of Lyon (1564) to Voltaire's 'Lettres philosophiques' (1734). 1984 ISBN 90-247-2923-8
105. R. Gascoigne: *Religion, Rationality and Community.* Sacred and Secular in the Thought of Hegel and His Critics. 1985 ISBN 90-247-2992-0
106. S. Tweyman: *Scepticism and Belief in Hume's 'Dialogues Concerning Natural Religion'.* 1986 ISBN 90-247-3090-2
107. G. Cerny: *Theology, Politics and Letters at the Crossroads of European Civilization.* Jacques Basnage [1653–1723] and the Baylean Huguenot Refugees in the Dutch Republic. 1987 ISBN 90-247-3150-X
108. Spinoza's *Algebraic Calculation of the Rainbow & Calculation of Changes.* Edited and Translated from Dutch, with an Introduction, Explanatory Notes and an Appendix by M.J. Petry. 1985 ISBN 90-247-3149-6
109. R.G. McRae: *Philosophy and the Absolute.* The Modes of Hegel's Speculation. 1985 ISBN 90-247-3151-8
110. J.D. North and J.J. Roche (eds.): *The Light of Nature.* Essays in the History and Philosophy of Science presented to A.C. Crombie. 1985 ISBN 90-247-3165-8
111. C. Walton and P.J. Johnson (eds.): *[Thomas] Hobbes's 'Science of Natural Justice'.* 1987 ISBN 90-247-3226-3
112. B.W. Head: *Ideology and Social Science.* Destutt de Tracy and French Liberalism. 1985 ISBN 90-247-3228-X
113. A.Th. Peperzak: *Philosophy and Politics.* A Commentary on the Preface to Hegel's *Philosophy of Right.* 1987 ISBN Hb 90-247-3337-5; Pb ISBN 90-247-3338-3

ARCHIVES INTERNATIONALES D'HISTOIRE DES IDÉES
*
INTERNATIONAL ARCHIVES OF THE HISTORY OF IDEAS

114. S. Pines and Y. Yovel (eds.): *Maimonides* [1135-1204] *and Philosophy.* Papers Presented at the 6th Jerusalem Philosophical Encounter (May 1985). 1986 ISBN 90-247-3439-8
115. T.J. Saxby: *The Quest for the New Jerusalem, Jean de Labadie* [1610–1674] *and the Labadists (1610–1744).* 1987 ISBN 90-247-3485-1
116. C.E. Harline: *Pamphlets, Printing, and Political Culture in the Early Dutch Republic.* 1987 ISBN 90-247-3511-4
117. R.A. Watson and J.E. Force (eds.): *The Sceptical Mode in Modern Philosophy.* Essays in Honor of Richard H. Popkin. 1988 ISBN 90-247-3584-X
118. R.T. Bienvenu and M. Feingold (eds.): *In the Presence of the Past.* Essays in Honor of Frank Manuel. 1991 ISBN 0-7923-1008-X
119. J. van den Berg and E.G.E. van der Wall (eds.): *Jewish-Christian Relations in the 17th Century.* Studies and Documents. 1988 ISBN 90-247-3617-X
120. N. Waszek: *The Scottish Enlightenment and Hegel's Account of 'Civil Society'.* 1988 ISBN 90-247-3596-3
121. J. Walker (ed.): *Thought and Faith in the Philosophy of Hegel.* 1991 ISBN 0-7923-1234-1
122. Henry More [1614–1687]: *The Immortality of the Soul.* Edited with Introduction and Notes by A. Jacob. 1987 ISBN 90-247-3512-2
123. P.B. Scheurer and G. Debrock (eds.): *Newton's Scientific and Philosophical Legacy.* 1988 ISBN 90-247-3723-0
124. D.R. Kelley and R.H. Popkin (eds.): *The Shapes of Knowledge from the Renaissance to the Enlightenment.* 1991 ISBN 0-7923-1259-7
125. R.M. Golden (ed.): *The Huguenot Connection.* The Edict of Nantes, Its Revocation, and Early French Migration to South Carolina. 1988 ISBN 90-247-3645-5
126. S. Lindroth: *Les chemins du savoir en Suède.* De la fondation de l'Université d'Upsal à Jacob Berzelius. Études et Portraits. Traduit du suédois, présenté et annoté par J.-F. Battail. Avec une introduction sur Sten Lindroth par G. Eriksson. 1988 ISBN 90-247-3579-3
127. S. Hutton (ed.): *Henry More (1614–1687). Tercentenary Studies.* With a Biography and Bibliography by R. Crocker. 1989 ISBN 0-7923-0095-5
128. Y. Yovel (ed.): *Kant's Practical Philosophy Reconsidered.* Papers Presented at the 7th Jerusalem Philosophical Encounter (December 1986). 1989 ISBN 0-7923-0405-5
129. J.E. Force and R.H. Popkin: *Essays on the Context, Nature, and Influence of Isaac Newton's Theology.* 1990 ISBN 0-7923-0583-3
130. N. Capaldi and D.W. Livingston (eds.): *Liberty in Hume's 'History of England'.* 1990 ISBN 0-7923-0650-3
131. W. Brand: *Hume's Theory of Moral Judgment.* A Study in the Unity of *A Treatise of Human Nature.* 1992 ISBN 0-7923-1415-8
132. C.E. Harline (ed.): *The Rhyme and Reason of Politics in Early Modern Europe.* Collected Essays of Herbert H. Rowen. 1992 ISBN 0-7923-1527-8
133. N. Malebranche: *Treatise on Ethics* (1684). Translated and edited by C. Walton. 1993 ISBN 0-7923-1763-7
134. B.C. Southgate: *'Covetous of Truth'.* The Life and Work of Thomas White (1593–1676). 1993 ISBN 0-7923-1926-5
135. G. Santinello, C.W.T. Blackwell and Ph. Weller (eds.): *Models of the History of Philosophy.* Vol. 1: From its Origins in the Renaissance to the 'Historia Philosophica'. 1993 ISBN 0-7923-2200-2
136. M.J. Petry (ed.): *Hegel and Newtonianism.* 1993 ISBN 0-7923-2202-9

ARCHIVES INTERNATIONALES D'HISTOIRE DES IDÉES
*
INTERNATIONAL ARCHIVES OF THE HISTORY OF IDEAS

137. Otto von Guericke: *The New (so-called Magdeburg) Experiments* [Experimenta Nova, Amsterdam 1672]. Translated and edited by M.G. Foley Ames. 1994 ISBN 0-7923-2399-8
138. R.H. Popkin and G.M. Weiner (eds.): *Jewish Christians and Cristian Jews*. From the Renaissance to the Enlightenment. 1994 ISBN 0-7923-2452-8
139. J.E. Force and R.H. Popkin (eds.): *The Books of Nature and Scripture*. Recent Essays on Natural Philosophy, Theology, and Biblical Criticism in the Netherlands of Spinoza's Time and the British Isles of Newton's Time. 1994 ISBN 0-7923-2467-6
140. P. Rattansi and A. Clericuzio (eds.): *Alchemy and Chemistry in the 16th and 17th Centuries*. 1994 ISBN 0-7923-2573-7
141. S. Jayne: *Plato in Renaissance England*. 1995 ISBN 0-7923-3060-9
142. A.P. Coudert: *Leibniz and the Kabbalah*. 1995 ISBN 0-7923-3114-1
143. M.H. Hoffheimer: *Eduard Gans and the Hegelian Philosophy of Law*. 1995 ISBN 0-7923-3114-1
144. J.R.M. Neto: *The Christianization of Pyrrhonism*. Scepticism and Faith in Pascal, Kierkegaard, and Shestov. 1995 ISBN 0-7923-3381-0
145. R.H. Popkin (ed.): *Scepticism in the History of Philosophy*. A Pan-American Dialogue. 1996 ISBN 0-7923-3769-7
146. M. de Baar, M. Löwensteyn, M. Monteiro and A.A. Sneller (eds.): *Choosing the Better Part*. Anna Maria van Schurman (1607–1678). 1995 ISBN 0-7923-3799-9
147. M. Degenaar: *Molyneux's Problem*. Three Centuries of Discussion on the Perception of Forms. 1996 ISBN 0-7923-3934-7
148. S. Berti, F. Charles-Daubert and R.H. Popkin (eds.): *Heterodoxy, Spinozism, and Free Thought in Early-Eighteenth-Century Europe*. Studies on the *Traité des trois imposteurs*. 1996 ISBN 0-7923-4192-9
149. G.K. Browning (ed.): *Hegel's* Phenomenology of Spirit: *A Reappraisal*. 1997 ISBN 0-7923-4480-4
150. G.A.J. Rogers, J.M. Vienne and Y.C. Zarka (eds.): *The Cambridge Platonists in Philosophical Context*. Politics, Metaphysics and Religion. 1997 ISBN 0-7923-4530-4
151. R.L. Williams: *The Letters of Dominique Chaix, Botanist-Curé*. 1997 ISBN 0-7923-4615-7
152. R.H. Popkin, E. de Olaso and G. Tonelli (eds.): *Scepticism in the Enlightenment*. 1997 ISBN 0-7923-4643-2
153. L. de la Forge. Translated and edited by D.M. Clarke: *Treatise on the Human Mind (1664)*. 1997 ISBN 0-7923-4778-1
154. S.P. Foster: *Melancholy Duty*. The Hume-Gibbon Attack on Christianity. 1997 ISBN 0-7923-4785-4
155. J. van der Zande and R.H. Popkin (eds.): *The Skeptical Tradition Around 1800*. Skepticism in Philosophy, Science, and Society. 1997 ISBN 0-7923-4846-X
156. P. Ferretti: *A Russian Advocate of Peace: Vasilii Malinovskii (1765–1814)*. 1997 ISBN 0-7923-4846-6
157. M. Goldish: *Judaism in the Theology of Sir Isaac Newton*. 1998 ISBN 0-7923-4996-2
158. A.P. Coudert, R.H. Popkin and G.M. Weiner (eds.): *Leibniz, Mysticism and Religion*. 1998 ISBN 0-7923-5223-8
159. B. Fridén: *Rousseau's Economic Philosophy*. Beyond the Market of Innocents. 1998 ISBN 0-7923-5270-X
160. C.F. Fowler O.P.: *Descartes on the Human Soul*. Philosophy and the Demands of Christian Doctrine. 1999 ISBN 0-7923-5473-7

ARCHIVES INTERNATIONALES D'HISTOIRE DES IDÉES
*
INTERNATIONAL ARCHIVES OF THE HISTORY OF IDEAS

161. J.E. Force and R.H. Popkin (eds.): *Newton and Religion.* Context, Nature and Influence. 1999 ISBN 0-7923-5744-2
162. J.V. Andreae: *Christianapolis.* Introduced and translated by E.H. Thompson. 1999 ISBN 0-7923-5745-0
163. A.P. Coudert, S. Hutton, R.H. Popkin and G.M. Weiner (eds.): *Judaeo-Christian Intellectual Culture in the Seventeeth Century.* A Celebration of the Library of Narcissus Marsh (1638–1713). 1999 ISBN 0-7923-5789-2
164. T. Verbeek (ed.): *Johannes Clauberg* and Cartesian Philosophy in the Seventeenth Century. 1999 ISBN 0-7923-5831-7
165. A. Fix: *Fallen Angels.* Balthasar Bekker, Spirit Belief, and Confessionalism in the Seventeenth Century Dutch Republic. 1999 ISBN 0-7923-5876-7
166. S. Brown (ed.): *The Young Leibniz and his Philosophy (1646–76).* 2000 ISBN 0-7923-5997-6
167. R. Ward: *The Life of Henry More.* Parts 1 and 2. 2000 ISBN 0-7923-6097-4
168. Z. Janowski: *Cartesian Theodicy.* Descartes' Quest for Certitude. 2000 ISBN 0-7923-6127-X
169. J.D. Popkin and R.H. Popkin (eds.): *The Abbé Grégoire and his World.* 2000 ISBN 0-7923-6247-0
170. C.G. Caffentzis: *Exciting the Industry of Mankind. George Berkeley's Philosophy of Money.* 2000 ISBN 0-7923-6297-7
171. A. Clericuzio: *Elements, Principles and Corpuscles.* A Study of Atomisms and Chemistry in the Seventeenth Century. 2000 ISBN 0-7923-6782-0
172. H. Hotson: *Paradise Postponed.* Johann Heinrich Alsted and the Birth of Calvinist Millenarianism. 2000 ISBN 0-7923-6787-1

KLUWER ACADEMIC PUBLISHERS – DORDRECHT / BOSTON / LONDON